T0127686

AN INTRODUCTION TO

BLACK HOLES, INFORMATION and the
STRING THEORY REVOLUTION

The Holographic Universe

LEONARD SUSSKIND • JAMES LINDESAY

Stanford University, USA   *Howard University, USA*

AN INTRODUCTION TO

BLACK HOLES, INFORMATION and the
STRING THEORY REVOLUTION

The Holographic Universe

World Scientific

NEW JERSEY • LONDON • SINGAPORE • BEIJING • SHANGHAI • HONG KONG • TAIPEI • CHENNAI

Published by

World Scientific Publishing Co. Pte. Ltd.

5 Toh Tuck Link, Singapore 596224

USA office: 27 Warren Street, Suite 401-402, Hackensack, NJ 07601

UK office: 57 Shelton Street, Covent Garden, London WC2H 9HE

British Library Cataloguing-in-Publication Data
A catalogue record for this book is available from the British Library.

AN INTRODUCTION TO BLACK HOLES, INFORMATION AND THE STRING THEORY REVOLUTION
The Holographic Universe

ISBN-13 978-981-256-083-4
ISBN-10 981-256-083-1
ISBN-13 978-981-256-131-2 (pbk)
ISBN-10 981-256-131-5 (pbk)

Printed in Singapore

About the Authors

Leonard Susskind is the Felix Bloch Professor of Theoretical Physics at Stanford University. He is a recipient of the J.J. Sakurai Prize and the Pregel Award, and is a member of the National Academy of Sciences and the American Academy of Arts and Sciences. He is also Distinguished Professor of Physics at the Korean Institute for Advanced Study, Extraordinary Professor of Physics at the University of Stellenbosch, as well as the Lorentz Professor for 2004 in the Netherlands.

James Lindesay is an Associate Professor of Physics at Howard University. His research and research training activities include work in theoretical and computational physics, quantum field theory and particle physics, condensed matter physics, atmospheric physics, astrophysics, biophysics and medical physics.

Preface

It is now almost a century since the year 1905, in which the principle of relativity and the hypothesis of the quantum of radiation were introduced. It has taken most of that time to synthesize the two into the modern quantum theory of fields and the standard model of particle phenomena. Although there is undoubtably more to be learned both theoretically and experimentally, it seems likely that we know most of the basic principles which follow from combining the special theory of relativity with quantum mechanics. It is unlikely that a major revolution will spring from this soil.

By contrast, in the 80 years that we have had the general theory of relativity, nothing comparable has been learned about the quantum theory of gravitation. The methods that were invented to quantize electrodynamics, which were so successfully generalized to build the standard model, prove wholly inadequate when applied to gravitation. The subject is riddled with paradox and contradiction. One has the distinct impression that we are thinking about the things in the wrong way. The paradigm of relativistic quantum field theory almost certainly has to be replaced.

How then are we to go about finding the right replacement? It seems very unlikely that the usual incremental increase of knowledge from a combination of theory and experiment will ever get us where we want to go, that is, to the Planck scale. Under this circumstance our best hope is an examination of fundamental principles, paradoxes and contradictions, and the study of gedanken experiments. Such strategy has worked in the past. The earliest origins of quantum mechanics were not experimental atomic physics, radioactivity, or spectral lines. The puzzle which started the whole thing was a contradiction between the principles of statistical thermodynamics and the field concept of Faraday and Maxwell. How was it possible, Planck asked, for the infinite collection of radiation oscillators to have a finite specific heat?

In the case of special relativity it was again a conceptual contradiction and a gedanken experiment which opened the way. According to Einstein, at the age of 15 he formulated the following paradox: suppose an observer moved along with a light beam and observed it. The electromagnetic field would be seen as a static, spatially varying field. But no such solution to Maxwell's equations exists. By this simple means a contradiction was exposed between the symmetries of Newton's and Galileo's mechanics and those of Maxwell's electrodynamics.

The development of the general theory from the principle of equivalence and the man-in-the-elevator gedanken experiment is also a matter of historical fact. In each of these cases the consistency of readily observed properties of nature which had been known for many years required revolutionary paradigm shifts.

What known properties of nature should we look to, and which paradox is best suited to our present purposes? Certainly the most important facts are the success of the general theory in describing gravity and of quantum mechanics in describing the microscopic world. Furthermore, the two theories appear to lead to a serious clash that once again involves statistical thermodynamics in an essential way. The paradox was discovered by Jacob Bekenstein and turned into a serious crisis by Stephen Hawking. By an analysis of gedanken experiments, Bekenstein realized that if the second law of thermodynamics was not to be violated in the presence of a black hole, the black hole must possess an intrinsic entropy. This in itself is a source of paradox. How and why a classical solution of field equations should be endowed with thermodynamical attributes has remained obscure since Bekenstein's discovery in 1972.

Hawking added to the puzzle when he discovered that a black hole will radiate away its energy in the form of Planckian black body radiation. Eventually the black hole must completely evaporate. Hawking then raised the question of what becomes of the quantum correlations between matter outside the black hole and matter that disappears behind the horizon. As long as the black hole is present, one can do the bookkeeping so that it is the black hole itself which is correlated to the matter outside. But eventually the black hole will evaporate. Hawking then made arguments that there is no way, consistent with causality, for the correlations to be carried by the outgoing evaporation products. Thus, according to Hawking, the existence of black holes inevitably causes a loss of quantum coherence and breakdown of one of the basic principles of quantum mechanics – the evolution of pure states to pure states. For two decades this contradiction between the principles of general relativity and quantum mechanics has so puzzled theorists that many now see it as a serious crisis.

Hawking and much of the traditional relativity community have been of the opinion that the correct resolution of the paradox is simply that quantum coherence is lost during black hole evaporation. From an operational viewpoint this would mean that the standard rules of quantum mechanics would not apply to processes involving black holes. Hawking further argued that once the loss of quantum coherence is permitted in black hole evaporation, it becomes compulsory in all processes involving the Planck scale. The world would behave as if it were in a noisy environment which continuously leads to a loss of coherence. The trouble with this is that there is no known way to destroy coherence without, at the same time violating energy conservation by heating the world. The theory is out of control as argued by Banks, Peskin and Susskind, and 't Hooft. Throughout this period, a few theorists, including 't Hooft and Susskind, have felt that the basic principles of quantum mechanics and statistical mechanics have to be made to co-exist with black hole evaporation.

't Hooft has argued that by resolving the paradox and removing the contradiction, the way to the new paradigm will be opened. The main purpose of this book is to lay out this case.

A second purpose involves development of string theory as a unified description of elementary particles, including their gravitational interactions. Although still very incomplete, string theory appears to be a far more consistent mathematical framework for quantum gravity than ordinary field theory. It is therefore worth exploring the differences between string theory and field theory in the context of black hole paradoxes. Quite apart from the question of the ultimate correctness and consistency of string theory, there are important lessons to be drawn from the differences between these two theories. As we shall see, although string theory is usually well approximated by local quantum field theory, in the neighborhood of a black hole horizon the differences become extreme. The analysis of these differences suggests a resolution of the black hole dilemma and a completely new view of the relations between space, time, matter, and information.

The quantum theory of black holes, with or without strings, is far from being a textbook subject with well defined rules. To borrow words from Sidney Coleman, it is a "trackless swamp" with many false but seductive paths and no maps. To navigate it without disaster we will need some beacons in the form of trusted principles that we can turn to for direction. In this book the absolute truth of the following four propositions will be assumed: 1) The formation and evaporation of a black hole is consistent with the basic principles of quantum mechanics. In particular, this means that observations performed by observers who remain outside the black hole can be described by a unitary time evolution. The global process, beginning with asymptotic infalling objects and ending with asymptotic outgoing evaporation products is consistent with the existence of a unitary S-matrix. 2) The usual semiclassical description of quantum fields in a slowly varying gravitational background is a good approximation to certain coarse grained features of the black hole evolution. Those features include the thermodynamic properties, luminosity, energy momentum flux, and approximate black body character of Hawking radiation. 3) Thirdly we assume the usual connection between thermodynamics and quantum statistical mechanics. Thermodynamics results from coarse graining a more microscopic description so that states with similar macroscopic behavior are lumped into a single thermodynamic state. The existence of a thermodynamics will be taken to mean that a microscopic set of degrees of freedom exists whose coarse graining leads to the thermal description. More specifically we assume that a thermodynamic entropy S implies that approximately $exp(S)$ quantum states have been lumped into one thermal state.

These three propositions, taken by themselves, are in no way radical. Proposition 1 and 3 apply to all known forms of matter. Proposition 2 may perhaps be less obvious, but it nevertheless rests on well-established foundations. Once we admit that a black hole has energy, entropy, and temperature, it must also have a luminosity. Furthermore the existence of a thermal behavior in the vicinity of the horizon follows from the equivalence principle as shown in the fundamental paper of Unruh. Why then should any of these principles be considered controversial? The answer lies in a fourth proposition which seems as inevitable as the first three: 4)   The fourth principle involves observers who fall through the horizon of a large massive black hole, carrying their laboratories with them. If the horizon scale is large enough so that tidal forces can be ignored, then a freely falling observer should detect nothing out of the ordinary when passing the horizon. The usual laws of nature with no abrupt external perturbations will be found valid until the influence of the singularity is encountered. In considering the validity of this fourth proposition it is important to keep in mind that the horizon is a global concept. The existence, location, size, and shape of a horizon depend not only on past occurrences, but also on future events. We ourselves could right now be at the horizon of a gigantic black hole caused by matter yet to collapse in the future. The horizon in classical relativity is simply the mathematical surface which separates those points from which any light ray must hit a singularity from those where light may escape to infinity. A mathematical surface of this sort should have no local effect on matter in its vicinity.

In Chapter 9 we will encounter powerful arguments against the mutual consistency of propositions 1–4. The true path through the swamp at times becomes so narrow it seems to be a dead end, while all around false paths beckon. Beware the will-o'-the-wisp and don't lose your nerve.

Contents

PART 1
Black Holes and Quantum Mechanics

Chapter 1

The Schwarzschild Black Hole

Before beginning the study of the quantum theory of black holes, one must first become thoroughly familiar with the geometry of classical black holes in a variety of different coordinate systems. Each coordinate system that we will study has its own particular utility, and no one of them is in any sense the best or most correct description. For example, the Kruskal–Szekeres coordinate system is valuable for obtaining a global overview of the entire geometry. It can however be misleading when applied to observations made by distant observers who remain outside the horizon during the entire history of the black hole. For these purposes, Schwarzschild coordinates, or the related tortoise coordinates, which cover only the exterior of the horizon are in many ways more valuable.

We begin with the simplest spherically symmetric static uncharged black holes described by Schwarzschild geometry.

1.1 Schwarzschild Coordinates

In Schwarzschild coordinates, the Schwarzschild geometry is manifestly spherically symmetric and static. The metric is given by

$$d\tau^2 = (1 - \tfrac{2MG}{r})dt^2 - (1 - \tfrac{2MG}{r})^{-1}dr^2 - r^2 d\Omega^2$$

$$= g_{\mu\nu}dx^\mu dx^\nu.$$

(1.1.1)

where $d\Omega^2 \equiv d\theta^2 + sin^2\theta d\phi^2$.

The coordinate t is called Schwarzschild time, and it represents the time recorded by a standard clock at rest at spatial infinity. The coordinate r is called the Schwarzschild radial coordinate. It does not measure proper spatial distance from the origin, but is defined so that the area of the 2-sphere at r is $4\pi r^2$. The angles θ, ϕ are the usual polar and azimuthal angles. In equation 1.1.1 we have chosen units such that the speed of light is 1.

The horizon, which we will tentatively define as the place where g_{00} vanishes, is given by the coordinate $r = 2MG$. At the horizon g_{rr} becomes singular. The question of whether the geometry is truly singular at the horizon or if it is the choice of coordinates which are pathological is subtle. In what follows we will see that no local invariant properties of the geometry are singular at $r = 2MG$. Thus a small laboratory in free fall at $r = 2MG$ would record nothing unusual. Nevertheless there is a very important sense in which the horizon is **globally** special if not singular. To a distant observer the horizon represents the boundary of the world, or at least that part which can influence his detectors.

To determine whether the local geometry is singular at $r = 2MG$ we can send an explorer in from far away to chart it. For simplicity let's consider a radially freely falling observer who is dropped from rest from the point $r = R$. The trajectory of the observer in parametric form is given by

$$r = \frac{R}{2}(1 + cos\eta) \tag{1.1.2}$$

$$\tau = \frac{R}{2}\left(\frac{R}{2MG}\right)^{1/2}(\eta + sin\eta) \tag{1.1.3}$$

$$t = (\tfrac{R}{2} + 2MG)\left(\tfrac{R}{2MG} - 1\right)^{1/2}\eta + \tfrac{R}{2}\left(\tfrac{R}{2MG} - 1\right)^{1/2}sin\eta$$
$$+2MG\,log\left|\frac{\left(\frac{R}{2MG}-1\right)^{1/2} + tan\frac{\eta}{2}}{\left(\frac{R}{2MG}-1\right)^{1/2} - tan\frac{\eta}{2}}\right|\qquad [0 < \eta < \pi] \tag{1.1.4}$$

where τ is the proper time recorded by the observer's clock. From these overly complicated equations it is not too difficult to see that the observer arrives at the point $r = 0$ after a finite interval

$$\tau = \frac{\pi}{2}R\left(\frac{R}{2MG}\right)^{\frac{1}{2}} \tag{1.1.5}$$

Evidently the proper time when crossing the horizon is finite and smaller than the expression in equation 1.1.5.

What does the observer encounter at the horizon? An observer in free fall is not sensitive to the components of the metric, but rather senses the tidal forces or curvature components. Define an orthonormal frame such that the observer is momentarily at rest. We can construct unit basis vectors, $\hat{\tau}$, $\hat{\rho}$, $\hat{\theta}$, $\hat{\phi}$ with $\hat{\tau}$ oriented along the observer's instantaneous time axis, and $\hat{\rho}$ pointing radially out. The non-vanishing curvature components are given by

$$R_{\hat{\tau}\hat{\theta}\hat{\tau}\hat{\theta}} = R_{\hat{\tau}\hat{\phi}\hat{\tau}\hat{\phi}} = -R_{\hat{\rho}\hat{\theta}\hat{\rho}\hat{\theta}} = -R_{\hat{\rho}\hat{\phi}\hat{\rho}\hat{\phi}} = \frac{MG}{r^3}$$

$$R_{\hat{\theta}\hat{\phi}\hat{\theta}\hat{\tau}} = -R_{\hat{\tau}\hat{\rho}\hat{\tau}\hat{\rho}} = \frac{2MG}{r^3}$$

(1.1.6)

Thus all the curvature components are finite and of order

$$R(Horizon) \sim \frac{1}{M^2 G^2}$$

(1.1.7)

at the horizon. For a large mass black hole they are typically very small. Thus the infalling observer passes smoothly and safely through the horizon.

On the other hand the tidal forces diverge as $r \to 0$ where a true local singularity occurs. At this point the curvature increases to the point where the classical laws of nature must fail.

Let us now consider the history of the infalling observer from the viewpoint of a distant observer. We may suppose that the infalling observer sends out signals which are received by the distant observer. The first surprising thing we learn from equations 1.1.2, 1.1.3, and 1.1.4 is that the crossing of the horizon does not occur at any finite Schwarzschild time. It is easily seen that as r tends to $2MG$, t tends to infinity. Furthermore a signal originating at the horizon cannot reach any point $r > 2MG$ until an infinite Schwarzschild time has elapsed. This is shown in Figure 1.1. Assuming that the infalling observer sends signals at a given frequency ν, the distant observer sees those signals with a progressively decreasing frequency. Over the entire span of Schwarzschild time the distant observer records only a finite number of pulses from the infalling transmitter. Unless the infalling observer increases the frequency of his/her signals to infinity as the horizon is approached, the distant observer will inevitably run out of signals and lose track of the transmitter after a finite number of pulses. The limits imposed on the information that can be transmitted from near the horizon are not so severe in classical physics as they are in quantum theory. According to classical physics the infalling observer can use an arbitrarily large carrier frequency to send an arbitrarily large amount of information using

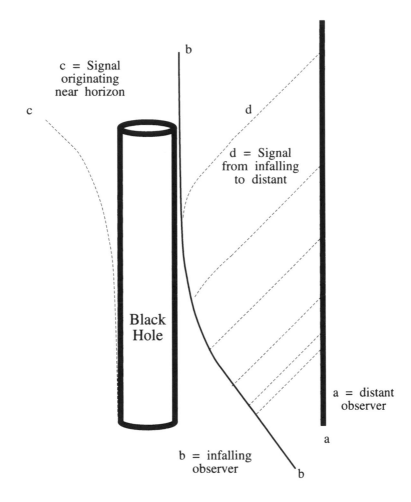

Fig. 1.1   *Infalling observer sending signals to distant Schwarzschild observer*

an arbitrarily small energy without significantly disturbing the black hole
and its geometry. Therefore, in principle, the distant observer can obtain
information about the neighborhood of the horizon and the infalling system right up to the point of horizon crossing. However quantum mechanics
requires that to send even a single bit of information requires a quantum of
energy. As the observer approaches the horizon, this quantum must have
higher and higher frequency, implying that the observer must have had a
large energy available. This energy will back react on the geometry, dis-

turbing the very quantity to be measured. Thereafter, as we shall see, no information can be transmitted from behind the horizon.

1.2 Tortoise Coordinates

A change of radial coordinate maps the horizon to minus infinity so that the resulting coordinate system covers only the region $r > 2MG$. We define the tortoise coordinate r^* by

$$\frac{1}{1 - \frac{2MG}{r}} dr^2 = \left(1 - \frac{2MG}{r}\right)(dr^*)^2 \tag{1.2.8}$$

so that

$$d\tau^2 = \left(1 - \frac{2MG}{r}\right)[dt^2 - (dr^*)^2] - r^2 d\Omega^2 \tag{1.2.9}$$

The interesting point is that the radial-time part of the metric now has a particularly simple form, called *conformally flat*. A space is called conformally flat if its metric can be brought to the form

$$d\tau^2 = F(x)\, dx^\mu dx^\nu\, \eta_{\mu\nu} \tag{1.2.10}$$

with $\eta_{\mu\nu}$ being the usual Minkowski metric. Any two-dimensional space is conformally flat, and a slice through Schwarzschild space at fixed θ, ϕ is no exception. In equation 1.2.9 the metric of such a slice is manifestly conformally flat. Furthermore it is also static.

The tortoise coordinate r^* is given explicitly by

$$r^* = r + 2MG\log\left(\frac{r - 2MG}{2MG}\right) \tag{1.2.11}$$

Note: $r^* \to -\infty$ at the horizon.
We shall see that wave equations in the black hole background have a very simple form in tortoise coordinates.

1.3   Near Horizon Coordinates (Rindler space)

The region near the horizon can be explored by replacing r by a coordinate ρ which measures proper distance from the horizon:

$$\rho = \int_{2MG}^{r} \sqrt{g_{rr}(r')}\, dr'$$

$$= \int_{2MG}^{r} (1 - \tfrac{2MG}{r'})^{-\frac{1}{2}}\, dr' \tag{1.3.12}$$

$$= \sqrt{r\,(r - 2MG)} + 2MG\, sinh^{-1}(\sqrt{\tfrac{r}{2MG} - 1})$$

In terms of ρ and t the metric takes the form

$$d\tau^2 = \left(1 - \frac{2MG}{r(\rho)}\right) dt^2 - d\rho^2 - r(\rho)^2\, d\Omega^2 \tag{1.3.13}$$

Near the horizon equation 1.3.12 behaves like

$$\rho \approx 2\sqrt{2MG(r - 2MG)} \tag{1.3.14}$$

giving

$$d\tau^2 \cong \rho^2 \left(\frac{dt}{4MG}\right)^2 - d\rho^2 - r^2(\rho)\, d\Omega^2 \tag{1.3.15}$$

Furthermore, if we are interested in a small angular region of the horizon arbitrarily centered at $\theta = 0$ we can replace the angular coordinates by Cartesian coordinates

$$x = 2MG\, \theta\, cos\phi$$

$$y = 2MG\, \theta\, sin\phi \tag{1.3.16}$$

Finally, we can introduce a dimensionless time ω

$$\omega = \frac{t}{4MG} \tag{1.3.17}$$

and the metric then takes the form

$$d\tau^2 = \rho^2\, d\omega^2 - d\rho^2 - dx^2 - dy^2 \tag{1.3.18}$$

It is now evident that ρ and ω are radial and hyperbolic angle variables for an ordinary Minkowski space. Minkowski coordinates T, Z can be defined by

$$T = \rho \sinh\omega$$

$$(1.3.19)$$

$$Z = \rho \cosh\omega$$

to get the familiar Minkowski metric

$$d\tau^2 = dT^2 - dZ^2 - dX^2 - dY^2 \qquad (1.3.20)$$

It should be kept in mind that equation 1.3.20 is only accurate near $r = 2MG$, and only for a small angular region. However it clearly demonstrates that the horizon is locally nonsingular, and, for a large black hole, is locally almost indistinguishable from flat space-time.

In Figure 1.2 the relation between Minkowski coordinates and the ρ, ω coordinates is shown. The entire Minkowski space is divided into four quadrants labeled I, II, III, and IV. Only one of those regions, namely

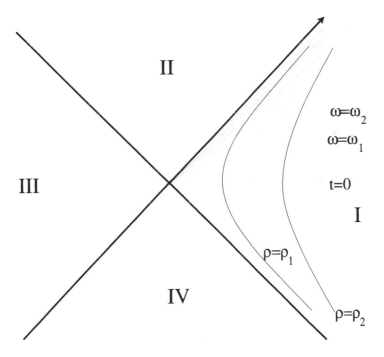

Fig. 1.2   *Relation between Minkowski and Rindler coordinates*

Region I lies outside the black hole horizon. The horizon itself is the origin $T = Z = 0$. Note that it is a two-dimensional surface in the four-dimensional space-time. This may appear surprising, since originally the horizon was defined by the single constraint $r = 2MG$, and therefore appears to be a three dimensional surface. However, recall that at the horizon g_{00} vanishes. Therefore the horizon has no extension or metrical size in the time direction.

The approximation of the near-horizon region by Minkowski space is called the Rindler approximation. In particular the portion of Minkowski space approximating the exterior region of the black hole, i.e. Region I, is called Rindler space. The time-like coordinate, ω, is called Rindler time. Note that a translation of Rindler time $\omega \to \omega + constant$ is equivalent to a Lorentz boost in Minkowski space.

1.4   Kruskal–Szekeres Coordinates

Finally we can bring the black hole metric to the form

$$d\tau^2 \;=\; F(R)\,[R^2\,d\omega^2 \;-\; dR^2] \;-\; r^2\,d\Omega^2 \qquad (1.4.21)$$

For small ρ equation 1.3.15 shows that $\rho \approx R$. A more accurate comparison with the original Schwarzschild metric gives the following requirements:

$$R^2\,F(R) \;=\; 16M^2G^2\left[1 - \frac{2MG}{r}\right] \qquad (1.4.22)$$

$$F(R)\,dR^2 \;=\; \frac{1}{1 - \frac{2MG}{r}}\,dr^2 \qquad (1.4.23)$$

from which it follows that

$$4MG\,log\frac{R}{MG} \;=\; r + 2MG\,log\left(\frac{r - 2MG}{2MG}\right) \;=\; r^* \qquad (1.4.24)$$

or

$$R \;=\; MG\,exp\left(\frac{r^*}{4MG}\right) \qquad (1.4.25)$$

R and ω can be thought of as radial and hyperbolic angular coordinates of a space which is conformal to flat 1+1 dimensional Minkowski space.

Letting

$$R\,e^{\omega} \;=\; V$$

$$R\,e^{-\omega} \;=\; -U$$

(1.4.26)

be "radial light-like" variables, the radial-time part of the metric takes the form

$$d\tau^2 \;=\; F(R)\,dU\,dV \tag{1.4.27}$$

The coordinates U, V are shown in Figure 1.3. The surfaces of constant r are the timelike hyperbolas in Figure 1.3. As r tends to $2MG$ the hyperbolas become the broken straight lines H^+ and H^- which we will call the extended past and future horizons. Although the extended horizons lie at finite values of the Kruskal–Szekeres coordinates, they are located at Schwarzschild time $\pm\infty$. Thus we see that a particle trajectory which crosses H^+ in a finite proper time, crosses $r = 2MG$ only after an infinite Schwarzschild time.

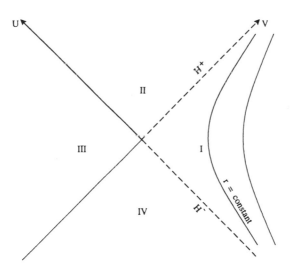

Fig. 1.3   *U,V Kruskal–Szekeres coordinates*

The region of Schwarzschild space with $r < 2MG$ can be taken to be Region II. In this region the surfaces of constant r are the

spacelike hyperboloids

$$UV = positive\ constant \qquad (1.4.28)$$

The true singularity at $r = 0$ occurs at $R^2 = -(MG)^2$, or

$$UV = (MG)^2 \qquad (1.4.29)$$

The entire maximal analytic extension of the Schwarzschild geometry is easily described in Kruskal–Szekeres coordinates. It is shown in Figure 1.4.

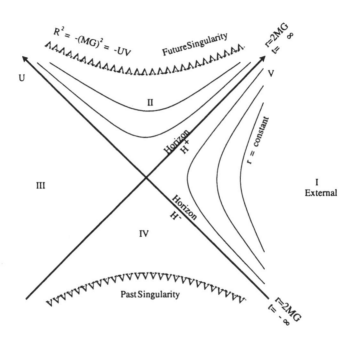

Fig. 1.4   *Maximal analytic extension of Schwarzschild in Kruskal–Szekeres coordinates*

A useful property of Kruskal–Szekeres coordinates is the fact that light rays and timelike trajectories always lie within a two-dimensional light cone bounded by 45° lines. A radial moving light ray travels on a trajectory $V = $ constant or $U = $ constant. A nonradially directed light ray or timelike trajectory always lies inside the two-dimensional light cone. With this in mind, it is easy to understand the causal properties of the black hole geometry. Consider a point P_1 in Region I. A radially outgoing light ray from P_1 will escape falling into the singularity as shown in Figure 1.5. An incoming light ray from P_1 will eventually cross H^+ and then hit the future singularity. Thus an observer in Region I can send messages to infinity as well as into Region II.

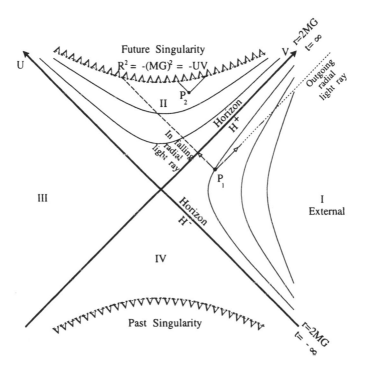

Fig. 1.5 *Radial light rays from a point in Region I and Region II using Kruskal–Szekeres coordinates*

Consider next Region II. From any point P_2 any signal must eventually hit the singularity. Furthermore, no signal can ever escape to Region I. Thus no observer who stays outside $r = 2MG$ can ever be influenced by events in Region II. For this reason Region II is said to be behind the horizon. Regions III and IV, as we will see, are not relevant to the classical problem of black holes formed by collapsing matter. Nevertheless let us consider them. From Region III no signal can ever get to Region I, and so it is also behind the horizon. On the other hand, points in Region IV can communicate with Region I. Region I however cannot communicate with Region IV. All of this is usually described by saying that Regions II and III are behind the future horizon while Regions III and IV are behind the past horizon.

1.5   Penrose Diagrams

Penrose diagrams are a useful way to represent the causal structure of spacetimes, especially if, like the Schwarzschild black hole, they have spherical symmetry. They represent the geometry of a two-dimensional surface of fixed angular coordinates. Furthermore they "compactify" the geometry so that it can be drawn in total on the finite plane. As an example, consider ordinary flat Minkowski space. Ignoring angular coordinates,

$$d\tau^2 = dt^2 - dr^2 - angular\,part = (dt + dr)(dt - dr) - angular\,part$$
$$(1.5.30)$$

Radial light rays propagate on the light cone $dt \pm dr = 0$.

Any transformation that is of the form

$$Y^+ = F(t + r)$$
$$Y^- = F(t - r)$$
$$(1.5.31)$$

will preserve the form of the light cone. We can use such a transformation to map the entire infinite space $0 \le r \le \infty$, $-\infty \le t \le +\infty$ to a finite portion of the plane. For example

$$Y^+ = tanh(t + r)$$
$$Y^- = tanh(t - r)$$
$$(1.5.32)$$

The entire space-time is mapped to the finite triangle bounded by

$$Y^+ = 1$$
$$Y^- = -1$$
$$Y^+ - Y^- = 0$$
$$(1.5.33)$$

as shown in Figure 1.6. Also shown in Figure 1.6 are some representative contours of constant r and t.

There are several infinities on the Penrose diagram. Future and past time-like infinities ($t = \pm\infty$) are the beginnings and ends of time-like trajectories. Space-like infinity ($r = \infty$) is where all space-like surfaces end. In addition to these there are two other infinities which are called $\mathcal{I}^\pm$.

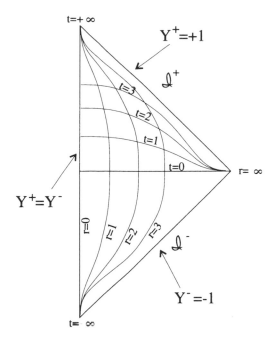

Fig. 1.6   *Penrose diagram for Minkowski space*

They are past and future light-like infinity, and they represent the origin
of incoming light rays and the end of outgoing light rays.

Similar deformations can be carried out for more interesting geometries,
such as the black hole geometry represent by Kruskal–Szekeres coordinates.
The resulting Penrose diagram is shown in Figure 1.7.

1.6   Formation of a Black Hole

The eternal black hole described by the static Schwarzschild geometry
is an idealization. In nature, black holes are formed from the collapse of
gravitating matter. The simpest model for black hole formation involves a
collapsing thin spherical shell of massless matter. For example, a shell of
photons, gravitons, or massless neutrinos with very small radial extension
and total energy M provides an example.

To construct the geometry, we begin with the empty space Penrose
diagram with the infalling shell represented by an incoming light-like line

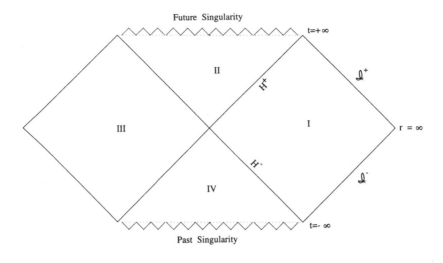

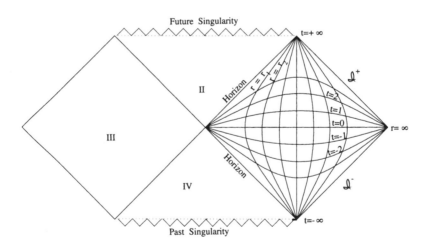

Fig. 1.7 *Penrose diagram for Schwarzschild black hole, showing regions (top) and curves of fixed radial position and constant time (bottom)*

(see Figure 1.8). The particular value of Y^+ chosen for the trajectory is arbitrary since any two such values are related by a time translation. The infalling shell divides the Penrose diagram into two regions, A and B. The Region A is interior to the shell and represents the initial flat space-time

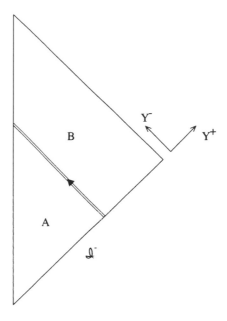

Fig. 1.8 *Minkowski space Penrose diagram for radially infalling spherical shell of massless particles with energy M*

before the shell passes. Region B is the region outside the shell and must be modified in order to account for the gravitational field due to the mass M.

In Newtonian physics the gravitational field exterior to a spherical mass distribution is uniquely that of a point mass located at the center of the distribution. Much the same is true in general relativity. In this case Birkoff's theorem tells us that the geometry outside the shell must be the Schwarzschild geometry. Accordingly, we consider the Penrose diagram for a black hole of mass M divided into regions A' and B' by an infalling massless shell as in Figure 1.9. Once again the particular value of Y^+ chosen for the trajectory is immaterial. Just as in Figure 1.8 where the Region B is unphysical, in Figure 1.9 the Region A' is to be discarded. To form the full classical evolution the regions A of Figure 1.8 and B' of Figure 1.9 must be glued together. However this must be done so that the "radius" of the local two sphere represented by the angular coordinates (θ, ϕ) is continuous. In other words, the mathematical identification of the boundaries of A and B' must respect the continuity of the variable r.

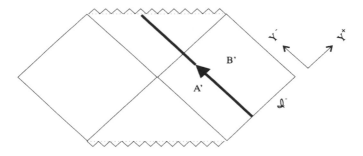

Fig. 1.9   *Penrose diagram of Schwarzschild black hole with radially infalling shell of massless particle with energy M*

Since in both cases r varies monotonically from $r = \infty$ at $\mathcal{I}^-$ to $r = 0$, the identification is always possible. One of the two Penrose diagrams will have to undergo a deformation along the Y^- direction in order to make the identification smoothly, but this will not disturb the form of the light cones. Thus in Figure 1.10 we show the resulting Penrose diagram for the complete geometry. On Fig 1.10, a light-like surface H is shown as a dotted line. It is clear that any light ray or timelike trajectory that originates to the upper left of H must end at the singularity and cannot escape to $\mathcal{I}^+$ (or $t = \infty$). This identifies H as the horizon. In Region B' the horizon is identical to the surface H^+ of Figure 1.7, that is it coincides with the future horizon of the final black hole geometry and is therefore found at $r = 2MG$. On the other hand, the horizon also extends into the Region A where the metric is just that of flat space-time. In this region the value of r on the horizon grows from an initial value $r = 0$ to the value $r = 2MG$ at the shell.

It is evident from this discussion that the horizon is a global and not a local concept. In the Region A no local quantity will distinguish the presence of the horizon whose occurence is due entirely to the future collapse of the shell.

Consider next a distant observer located on a trajectory with $r \gg 2MG$. The observer originates at past time-like infinity and eventually ends at future time-like infinity, as shown in Figure 1.11. The distant observer collects information that arrives at any instant from his backward light cone. Evidently such an observer never actually sees events on the

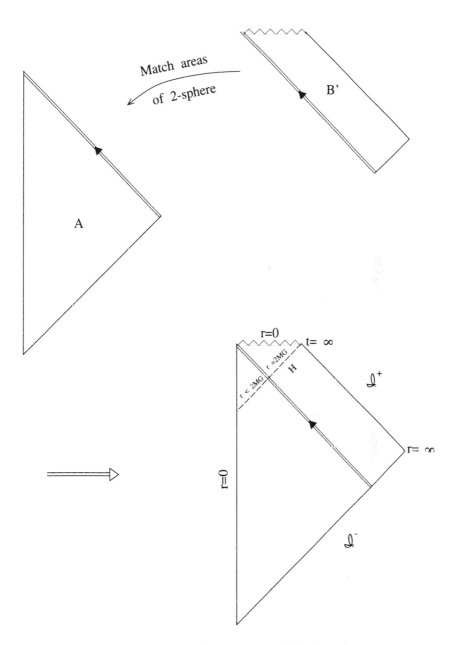

Fig. 1.10 *Penrose diagram for collapsing shell of massless particles*

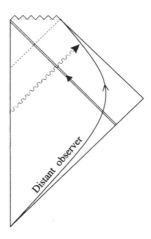

Fig. 1.11   *Distant observer to collapsing spherical shell*

horizon. In this sense the horizon must be regarded as at the end of time. Any particle or wave which falls through the horizon is seen by the distant observer as asymptotically approaching the horizon as it is infinitely red shifted. At least that is the case classically.

This basic description of black hole formation is much more general than might be guessed. It applies with very little modification to the collapse of all kinds of massive matter as well as to non-spherical distributions. In all cases the horizon is a lightlike surface which separates the space-time into an inner and an outer region. Any light ray which originates in the inner region can never reach future asymptotic infinity, or for that matter ever reach any point of the outer region. The events in the outer region can send light rays to $\mathcal{I}^+$ and time-like trajectories to $t = \infty$.

The horizon, as we have seen, is a global concept whose location depends on all future events. It is composed of a family of light rays or null geodesics, passing through each space-time point on the horizon. This is shown in Figure 1.12. Notice that null geodesics are vertical after the shell crosses the horizon and essentially at 45° prior to that crossing. These light rays are called the *generators* of the horizon.

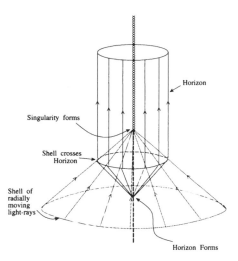

Fig. 1.12 *The horizon as family of null geodesics*

1.7 Fidos and Frefos and the Equivalence Principle

In considering the description of events near the horizon of a **static** black hole from the viewpoint of an external observer[1], it is useful to imagine space to be filled with static observers, each located at a fixed (r, θ, ϕ). Such observers are called fiducial observers, or by the whimsical abbreviation, FIDOS. Each Fido carries a clock which may be adjusted to record Schwarzschild time t. This means that Fidos at different r values see their own clocks running at different proper rates. Alternatively, they could carry standard clocks which always record proper time τ. At a given r the relation between Schwarzschild time t and the Fidos proper time τ is given by

$$\frac{d\tau}{dt} = \sqrt{g_{00}} = [1 - \frac{2MG}{r}]^{\frac{1}{2}} \qquad (1.7.34)$$

Thus, to the Fido near $r = 2MG$, the Schwarzschild clock appears to run at a very rapid rate. Another possible choice of clocks would record the dimensionless hyperbolic angle ω defined by equation 1.3.17.

The spatial location of the Fidos can be labeled by the angular coordinates (θ, ϕ) and any one of the radial variables r, r^*, or ρ. Classically the Fidos can be thought of as mathematical fictions or real but arbitrarily light systems suspended by arbitrarily light threads from some sort of suspension system built around the black hole at a great distance. The acceleration of a Fido at proper distance ρ is given by $\frac{1}{\rho}$ for $\rho << MG$. Quantum mechanically we have a dilemma if we try to imagine the Fidos as real. If they are extrememly light their locations will necessarily suffer large quantum fluctuations, and they will not be useful as fixed anchors labeling space-time points. If they are massive they will influence the gravitational field that we wish to describe. Quantum mechanically, physical Fidos must be replaced by a more abstract concept called *gauge fixing*. The concept of gauge fixing in gravitation theory implies a mathematical restriction on the choice of coordinates. However all real observables are required to be gauge invariant.

Now let us consider a classical particle falling radially into a black hole. There are two viewpoints we can adopt toward the description of the particle's motion. The first is the viewpoint of the Fidos who are permanently stationed outside the black hole. It is a viewpoint which is also useful to a distant observer, since any observation performed by a Fido can be communicated to distant observers. According to this viewpoint, the particle never crosses the horizon but asymptotically approaches it. The second viewpoint involves freely falling observers (FREFOS) who follow the particle as it falls. According to the Frefos, they and the particle cross the horizon after a finite time. However, once the horizon is crossed, their observations cannot be communicated to any Fido or to a distant observer.

Once the infalling particle is near the horizon its motion can be described by the coordinates (T, Z, X, Y) defined in equations 1.3.16 and 1.3.19. Since the particle is freely falling, in the Minkowski coordinates its motion is a straight line

$$\frac{dZ}{d\tau} = \frac{p^z}{m} = -\frac{p_Z}{m}$$
$$\frac{dT}{d\tau} = \frac{p_T}{m} \tag{1.7.35}$$

where p_Z and p_T are the Z and T components of momentum, and m is the mass of the particle. As the particle freely falls past the horizon, the components p_Z and p_T may be regarded as constant or slowly varying. They are the components seen by Frefos.

The components of momentum seen by Fidos are the components p_ρ and p_τ which, using equation 1.3.19, are given by

$$p_\rho = p_Z \cosh\omega + p_T \sinh\omega$$
$$p_\tau = p_Z \sinh\omega + p_T \cosh\omega \tag{1.7.36}$$

For large times we find

$$p_\rho \approx p_\tau \approx 2p_Z \, exp\omega = 2p_Z \, exp(\frac{t}{4MG}) \qquad (1.7.37)$$

Thus we find the momentum of an infalling particle as seen by a Fido grows exponentially with time! It is also easily seen that ρ, the proper spatial distance of the particle from the horizon, exponentially decreases with time

$$\rho(t) \approx \rho(0)exp(-\frac{t}{4MG}) \qquad (1.7.38)$$

Locally the relation between the coordinates of the Frefos and Fidos is a time dependent boost along the radial direction. The hyperbolic boost angle is the dimensionless time ω. Eventually, during the lifetime of the black hole this boost becomes so large that the momentum of an infalling particle (as seen by a Fido) quickly exceeds the entire mass of the universe.

As a consequence of the boost, the Fidos see all matter undergoing Lorentz contraction into a system of arbitrarily thin "pancakes" as it approaches the horizon. According to classical physics, the infalling matter is stored in "sedimentary" layers of diminishing thickness as it eternally sinks toward the horizon (see Figure 1.13). Quantum mechanically we must expect this picture to break down by the time the infalling particle has been squeezed to within a Planck distance from the horizon. The Frefos of course see the matter behaving in a totally unexceptional way.

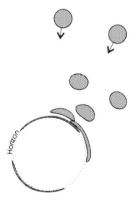

Fig. 1.13 *Sedimentary layers of infalling matter on horizon*

Chapter 2

Scalar Wave Equation in a Schwarzschild Background

In Chapters 3 and 4 we will be concerned with the behavior of quantum fields near horizons. In this lecture we will study the properties of a scalar wave equation in the background of a black hole.

Let us consider a conventional massless free Klein–Gordon field χ in the Schwarzschild background. Here we will find great advantage in utilizing tortoise coordinates in which the metric has the form

$$d\tau^2 = F(r^*)[dt^2 - (dr^*)^2] - r^2[d\theta^2 + sin^2\theta d\phi^2] \qquad (2.0.1)$$

The action for χ is

$$I = \tfrac{1}{2} \int \sqrt{-g}\, g^{\mu\nu}\, \partial_\mu\chi\, \partial_\nu\chi\, d^4x$$

$$= \tfrac{1}{2} \int dt\, dr^*\, d\theta\, d\phi\, \{ \tfrac{(\partial_t\chi)^2 - (\partial_{r^*}\chi)^2}{F} \qquad (2.0.2)$$

$$- \tfrac{1}{r^2}(\tfrac{\partial\chi}{\partial\theta})^2 - \tfrac{1}{r^2 sin^2\theta}(\tfrac{\partial\chi}{\partial\phi})^2 \} F r^2 sin\theta$$

Now define

$$\psi = r\chi \qquad (2.0.3)$$

and the action takes the form

$$I = \tfrac{1}{2} \int [(\partial_t\psi)^2 - (\tfrac{\partial\psi}{\partial r^*} - \tfrac{\partial(lnr)}{\partial r^*}\psi)^2$$

$$- \tfrac{F}{r^2}(sin\theta\left(\tfrac{\partial\psi}{\partial\theta}\right)^2 + \tfrac{1}{sin\theta}\left(\tfrac{\partial\psi}{\partial\phi}\right)^2)]\, dt\, dr^*\, d\theta\, d\phi \qquad (2.0.4)$$

25

which, after an integration by parts and the introduction of spherical harmonic decomposition becomes

$$I = \sum_{\ell m} \tfrac{1}{2} \int [(\dot{\psi}_{\ell m})^2 - \left(\frac{\partial \psi_{\ell m}}{\partial r^*}\right)^2 +$$

$$-\{\left(\frac{\partial \ln r}{\partial r^*}\right)^2 + \frac{\partial}{\partial r^*}\left(\frac{\partial \ln r}{\partial r^*}\right)\}\, \psi_{\ell m}^2 - \frac{F}{r^2}\ell(\ell+1)\,\psi_{\ell m}^2]dt\,dr^* \tag{2.0.5}$$

Using the relation between r and r^*

$$r^* = r + 2MG \ln(r - 2MG)$$

gives for each ℓ, m an action

$$I_{\ell m} = \frac{1}{2}\int dt\,dr^* \left[\left(\frac{\partial \psi_{\ell m}}{\partial t}\right)^2 - \left(\frac{\partial \psi_{\ell m}}{\partial r^*}\right)^2 - V_\ell(r^*)\,\psi_{\ell m}^2\right] \tag{2.0.6}$$

where the potential $V_\ell(r^*)$ is given by

$$V_\ell(r^*) = \frac{r - 2MG}{r}\left(\frac{\ell(\ell+1)}{r^2} + \frac{2MG}{r^3}\right) \tag{2.0.7}$$

The equation of motion is

$$\ddot{\psi}_{\ell m} = \frac{\partial^2 \psi_{\ell m}}{(\partial r^*)^2} - V_\ell(r^*)\,\psi_{\ell m} \tag{2.0.8}$$

and for a mode of frequency ν

$$-\frac{\partial^2 \psi_{\ell m}}{(\partial r^*)^2} + V_\ell(r^*)\,\psi_{\ell m} = \nu^2\,\psi_{\ell m} \tag{2.0.9}$$

The potential V is shown in Figure 2.1 as a function of the Schwarzschild coordinate r. For $r \gg 3MG$ the potential is repulsive. In fact it is just the relativistic generalization of the usual repulsive centrifugal barrier. However as the horizon is approached, gravitational attraction wins and the potential becomes attractive, and pulls a wave packet toward the horizon. The maximum of the potential, where the direction of the force changes, depends weakly on the angular momentum ℓ. It is given by

$$r_{max} = 3MG \left(\frac{1}{2}\left(1 + \sqrt{1 + \frac{14\ell^2 + 14\ell + 9}{9\ell^2(\ell+1)^2}}\right) - \frac{1}{2\ell(\ell+1)}\right) \tag{2.0.10}$$

For $\ell \to \infty$ the maximum occurs at $r_{max}(\ell \to \infty) = 3MG$.

The same potential governs the motion of massless classical particles. One can see that the points $r_{max}(\ell)$ represent unstable circular orbits, and

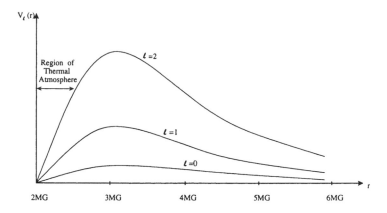

Fig. 2.1 *Effective potential for free scalar field vs Schwarzschild radial coordinate*

the innermost such orbit is at $r = 3MG$. Any particle that starts with vanishing radial velocity in the region $r < 3MG$ will spiral into the horizon.

In the region of large negative r^* where we approach the horizon, the potential is unimportant, and the field behaves like a free massless Klein–Gordon field. The eigenmodes in this region have the form of plane waves which propagate with unit velocity

$$\frac{dr^*}{dt} = \mp 1$$

$$\psi \to e^{i\,k\,(r^*\pm t)}$$

(2.0.11)

Let us consider a field quantum of frequency ν and angular momentum ℓ propagating from large negative r^* toward the barrier at $r \approx 3MG$. Will it pass over the barrier? To answer this we note that equation 2.0.9 has the form of a Schrödinger equation for a particle of energy ν^2 in a potential V. The particle has enough energy to overcome the barrier without tunneling if ν^2 is larger than the maximum height of the barrier. For example, if $\ell = 0$ the height of the barrier is

$$V_{max} = \frac{1}{2M^2G^2}\left(\frac{3}{8}\right)^3$$

(2.0.12)

An s-wave quantum will therefore escape if

$$\nu > \frac{0.15}{MG}$$

(2.0.13)

Similarly an s-wave quantum with $\nu > \frac{0.15}{MG}$ will be able to penetrate the barrier from the outside and fall to the horizon. Less energetic particles must tunnel through the barrier.

A particle of high angular momentum, whether on the inside or outside of the barrier will have more difficulty penetrating through. For large ℓ

$$V_{max} \approx \frac{1}{27} \frac{\ell^2}{M^2 G^2} \tag{2.0.14}$$

Therefore the threshold energy for passing over the barrier is

$$\nu \sim \frac{1}{\sqrt{27}} \frac{\ell}{MG} \tag{2.0.15}$$

2.1   Near the Horizon

Near the horizon the exterior of the black hole may be described by the Rindler metric

$$d\tau = \rho^2 \, d\omega^2 - d\rho^2 - dX^2 - dY^2$$

It is useful to replace ρ by a tortoise-like coordinate which again goes to $-\infty$ at the horizon. We define

$$u = log\rho \tag{2.1.16}$$

and the metric near the horizon becomes

$$d\tau^2 = exp(2u) \left[d\omega^2 - du^2 \right] - dX^2 - dY^2 \tag{2.1.17}$$

The scalar field action becomes

$$I = \frac{1}{2} \int dX \, dY \, du \, d\omega \left[\left(\frac{\partial \chi}{\partial \omega} \right)^2 - \left(\frac{\partial \chi}{\partial u} \right)^2 - e^{2u} \left(\partial_\perp \chi \right)^2 \right] \tag{2.1.18}$$

where $\partial_\perp \chi = (\partial_X, \partial_Y)$. Instead of using spherical waves, near the horizon we can decompose χ into transverse plane waves with transverse wave vector $k_\perp$

$$\chi = \int d^2 k_\perp \, e^{i k_\perp \, x_\perp} \, \chi(k_\perp, u, \omega) \tag{2.1.19}$$

the action for a given wave number k is

$$I = \frac{1}{2} \int d\omega \, du \, \left[(\partial_\omega \chi)^2 - (\partial_u \chi)^2 - k^2 \, e^{2u} \, \chi^2 \right] \qquad (2.1.20)$$

Thus the potential is

$$V(k, u) = k^2 \, e^{2u} \qquad (2.1.21)$$

The correspondence between the momentum vector k and the angular momentum ℓ is given by the usual connection between momentum and angular momentum. If the horizon has circumference $2\pi(2MG)$, then a wave with wave vector $k_\perp$ will correspond to an angular momentum $|\ell| = |k| \, r = 2MG|k|$. Thus the potential in equation 2.1.21 is seen to be proportional to ℓ^2. For very low angular momentum the approximation is not accurate, but qualitatively is correct for $\ell > 0$. In approximating a sum over ℓ and m by an integral over k, the integral should be infrared cut off at $|k| \sim \frac{1}{MG}$.

From the action in equation 2.1.20 we obtain the equation of motion

$$\frac{\partial^2 \chi}{\partial \omega^2} - \frac{\partial^2 \chi}{\partial u^2} + k^2 \, exp(2u) \, \chi = 0 \qquad (2.1.22)$$

A solution which behaves like $exp(i\nu t)$ in Schwarzschild time has the form

$$e^{i \nu \, [4MG\omega]} - e^{i \lambda \omega} \qquad (2.1.23)$$

The time independent form of the equation of motion is

$$-\frac{\partial^2 \chi}{\partial u^2} + \left(k^2 \, exp(2u) \right) \chi = \lambda^2 \chi \qquad (2.1.24)$$

Once again we see that unless $k = 0$, there is a potential confining quanta to the region near the horizon. Qualitatively, the behavior of a quantum field in a black hole background differs from the Rindler space approximation in that for the black hole, the potential barrier is cut off when $\rho = e^u$ is greater than MG. By contrast, in the Rindler case V increases as e^u without bound.

We have not thus far paid attention to the boundary conditions at the horizon where $u \to -\infty$. Since in this region the field $\chi(u)$ behaves like a free massless field, the boundary condition would be expected to be that the field is in the usual quantum ground state. In the next section we will see that this is not so.

Chapter 3

Quantum Fields in Rindler Space

According to Einstein, the study of a phenomenon in a gravitational field is best preceeded by a study of the same phenomenon in an accelerated coordinate system. In that way we can use the special relativistic laws of nature to understand the effect of a gravitational field.

As we have seen, the relativistic analogue of a uniformly accelerated frame is Rindler space. Because Rindler space covers only a portion of the space time geometry (Region I) there are new and subtle features to the description of quantum fields. These features are closely associated with the existence of the horizon. The method we will use applies to any relativistic quantum field theory including those with nontrivial interactions. For illustrative purposes we will consider a free scalar field theory. It is important to bare in mind that such a non-interacting description is of limited validity. As we shall see, interactions become very important near the horizon of a black hole. Ignoring them leads to an inconsistent description of the Hawking evaporation process.

3.1 Classical Fields

First let us consider the evolution of a classical field in Rindler space. The field in Region I of Figure 1.2 can be described in a self contained way. Obviously influences from Regions II and III can never be felt in Region I since no point in Regions II or III is in the causal past of any point in Region I. Signals from Region IV can, of course, reach Region I, but to do so they must pass through the surface $\omega = -\infty$. Therefore signals from Region IV are regarded as initial data in the remote past by the Rindler observer. Evidently the Rindler observer sees a world in which physical phenomena can be described in a completely self contained way.

The evolution from one surface of constant ω to another is governed by the Rindler Hamiltonian. Using conventional methods the generator of ω-translations is given by

$$H_R = \int_{\rho=0}^{\infty} d\rho \, dX \, dY \, \rho \, T^{00}(\rho, X, Y) \qquad (3.1.1)$$

where T^{00} is the usual Hamiltonian density used by the Minkowski observer. For example, for a massive scalar field with potential V, T^{00} is given by

$$T^{00} = \frac{\Pi^2}{2} + \frac{1}{2} (\nabla \chi)^2 + V(\chi) \qquad (3.1.2)$$

where Π is the canonical momentum conjugate to χ. The Rindler Hamiltonian is

$$H_R = \int d\rho \, dx_\perp \frac{\rho}{2} \left[\Pi^2 + \left(\frac{\partial \chi}{\partial \rho} \right)^2 + \left(\frac{\partial \chi}{\partial x_\perp} \right)^2 + 2V(\chi) \right] \qquad (3.1.3)$$

The origin of the factor ρ in the Rindler Hamiltonian density is straightforward. In Figure 3.1 the relation between neighboring equal Rindler-time surfaces is shown. The proper time separation between the surfaces is

$$\delta\tau = \rho \, \delta\omega \qquad (3.1.4)$$

Thus, to push the ω-surface ahead requires a ρ-dependent time translation. This is the reason that T^{00} is weighted with the factor ρ. The Rindler Hamiltonian is similar to the generator of Lorentz boosts from the viewpoint of the Minkowski observer. However it only involves the degrees of freedom in Region I.

3.2   Entanglement

Quantum fields can also be described in a self-contained fashion in Rindler space, but a new twist is encountered. Our goal is to describe the usual physics of a quantum field in Minkowski space, but from the viewpoint of the Fidos in Region I, i.e. in Rindler space. To understand the new feature, recall that in the usual vacuum state, the correlation between fields at different spatial points does not vanish. For example, in free massless scalar theory the equal time correlator is given by

$$\langle 0 | \, \chi(X, Y, Z) \, \chi(X', Y', Z') \, | 0 \rangle \sim \frac{1}{\Delta^2} \qquad (3.2.5)$$

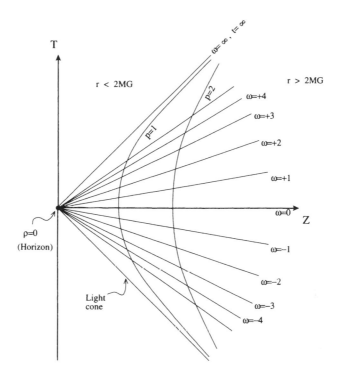

Fig. 3.1 *Equal time and proper distance surfaces in Rindler space*

where Δ is the space-like separation between the points (X, Y, Z) and (X', Y', Z')

$$\Delta^2 = (X - X')^2 + (Y - Y')^2 + (Z - Z')^2 \qquad (3.2.6)$$

The two points might both lie within Region I, in which case the correlator in equation 3.2.5 represents the quantum correlation seen by Fido's in Region I. On the other hand, the two points might lie on opposite sides of the horizon at $Z = 0$. In that case the correlation is unmeasurable to the Fidos in Region I. Nevertheless it has significance. When two subsystems (fields in Regions I and III) become correlated, we say that they are quantum entangled, so that neither can be described in terms of pure states. The appropriate description of an entangled subsystem is in terms of a density matrix.

3.3   Review of the Density Matrix

Suppose a system consists of two subsystems, A and B, which have previously been in contact but are no longer interacting. The combined system has a wavefunction

$$\Psi = \Psi(\alpha, \beta) \tag{3.3.7}$$

where α and β are appropriate commuting variables for the subsystems A and B.

Now suppose we are only interested in subsystem A. A complete description of all measurements of A is provided by the density matrix $\rho_A(\alpha, \alpha')$.

$$\rho_A(\alpha, \alpha') = \sum_\beta \Psi^*(\alpha, \beta)\,\Psi(\alpha', \beta) \tag{3.3.8}$$

Similarly, experiments performed on B are described by $\rho_B(\beta, \beta')$.

$$\rho_B(\beta, \beta') = \sum_\alpha \Psi^*(\alpha, \beta)\,\Psi(\alpha, \beta') \tag{3.3.9}$$

The rule for computing an expectation value of an operator $\mathbf{a}$ composed of A degrees of freedom is

$$\langle \mathbf{a} \rangle = Tr\,\mathbf{a}\,\rho_A \tag{3.3.10}$$

Density matrices have the following properties:
1) $Tr\,\rho = 1$ (total probability=1)
2) $\rho = \rho^\dagger$ (hermiticity)
3) $\rho_j \geq 0$ (all eigenvalues are positive or zero)

In the representation in which ρ is diagonal

$$\rho = \begin{pmatrix} \rho_1 & 0 & 0 & \dots & \dots \\ 0 & \rho_2 & 0 & \dots & \dots \\ 0 & 0 & \rho_3 & \dots & \dots \\ \dots & \dots & \dots & \dots & \dots \end{pmatrix} \tag{3.3.11}$$

The eigenvalues ρ_j can be considered to be probabilities that the system is in the jth state. However, unlike the case of a coherent superposition of states, the relative phases between the states $|j\rangle$ are random.

There is one special case when the density matrix is indistinguishable from a pure state. This is the case in which only one eigenvalue ρ_j is nonzero. This case can only result from an uncorrelated product wave function of the form

$$\Psi(\alpha, \beta) = \psi_A(\alpha) \psi_B(\beta) \tag{3.3.12}$$

A quantitative measure of the departure from a pure state is provided by the *Von Neumann entropy*

$$S = -Tr\rho \log \rho = -\sum_j \rho_j \log \rho_j. \tag{3.3.13}$$

S is zero if and only if all the eigenvalues but one are zero. The one non-vanishing eigenvalue is equal to 1 by virtue of the trace condition on ρ. The entropy is also a measure of the degree of entanglement between A and B. It is therefore called the *entropy of entanglement*.

The opposite extreme to a pure state is a completely incoherent density matrix in which all the eigenvalues are equal to $\frac{1}{N}$, where N is the dimensionality of the Hilbert space. In that case S takes its maximum value

$$S_{max} = -\sum_j \frac{1}{N} \log \frac{1}{N} = \log N \tag{3.3.14}$$

More generally, if ρ is a projection operator onto a subspace of dimension n, we find

$$S = \log n \tag{3.3.15}$$

Thus we see that the Von Neumann entropy is a measure of the number of states which have an appreciable probability in the statistical ensemble. We may think of e^S as an effective dimensionality of the subspace described by ρ.

The Von Neumann (or entanglement) entropy should not be confused with the thermal entropy of the second law of thermodynamics. This entropy has its origin in coarse graining. If a system with Hamiltonian H is in thermal equilibrium at temperature $T = 1/\beta$ then it is described by a Maxwell–Boltzman density matrix

$$\rho_{M.B.} = \frac{e^{-\beta H}}{Tr e^{-\beta H}}. \tag{3.3.16}$$

In this case the thermal entropy is given by

$$S_{thermal} = -Tr\, \rho_{M.B.}\, \log \rho_{M.B.} \tag{3.3.17}$$

3.4   The Unruh Density Matrix

Now let us consider the space of states describing a Lorentz invariant quantum field theory in Minkowski space. In Figure 3.2 the surface $T = 0$ of Minkowski space is shown divided into two halves, one in Region I and one in Region III. For the case of a scalar field χ the fields at each point

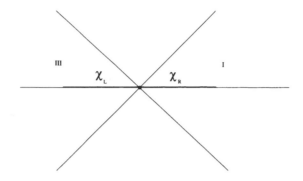

Fig. 3.2   *Fields on the spacelike surface $T = 0$ in Minkowski space*

of space form a complete set of commuting operators. These fields may be decomposed into two subsets associated with regions I and III. we call them χ_R and χ_L respectively. Thus

$$\chi(X,Y,Z) = \chi_R(X,Y,Z) \ Z > 0$$
$$\chi(X,Y,Z) = \chi_L(X,Y,Z) \ Z < 0$$

(3.4.18)

The general wave functional of the system is a functional of χ_L and χ_R

$$\Psi = \Psi(\chi_L, \chi_R)$$

(3.4.19)

We wish to compute the density matrix used by the Fidos in Region I to describe their Rindler world. In particular we would like to understand the density matrix ρ_R which represents the usual Minkowski vacuum to the Fidos in Region I.

First let us see what we can learn from general principles. Obviously the state Ψ is translationally invariant under the usual Minkowski space translations. Thus the Fidos must see the vacuum as invariant under translations along the X and Y axes. However, the translation invariance along the Z axis is explicitly broken by the act of singling out the origin $Z = 0$ for special consideration. From the X, Y translation invariance we conclude that ρ_R commutes with the components of momentum in these directions

$$[p_X, \rho_R] = [p_Y, \rho_R] = 0 \qquad (3.4.20)$$

A very important property of ρ_R is that it is invariant under Rindler time translations $\omega \to \omega + constant$. This follows from the Lorentz boost invariance of Ψ. Thus

$$[H_R, \rho_R] = 0 \qquad (3.4.21)$$

To proceed further we must use the fact that $\Psi(\chi_L, \chi_R)$ is the ground state of the Minkowski Hamiltonian. General path integral methods may be brought to bare on the computation of the ground state wave functional. Let us assume that the field theory is described in terms of an action

$$I = \int d^3 X \, dT \, L \qquad (3.4.22)$$

The so-called Euclidean field theory is defined by replacing the time coordinate T by $i X^0$. For example, the Euclidean version of ordinary scalar field theory is obtained from the usual Minkowski action

$$I = \int d^3 X \, dT \, \frac{1}{2} \left[\dot{\chi}^2 - (\nabla \chi)^2 - 2V(\chi) \right] \qquad (3.4.23)$$

Letting $T \to i X^0$ we obtain the Euclidean action

$$I_E = \int d^4 X \, \frac{1}{2} \left[(\partial_X \chi)^2 - 2V(\chi) \right] \qquad (3.4.24)$$

Now a standard method of computing the ground state by path integration is to use the Feynman–Hellman theorem. Suppose we wish to compute $\Psi(\chi_L, \chi_R)$. Then we consider the path integral

$$\Psi(\chi_L, \chi_R) = \frac{1}{\sqrt{Z}} \int d\chi(x) \, e^{-I_E} \qquad (3.4.25)$$

where the path integral is over all $\chi(x)$ with $X^0 > 0$ and Z is an appropriate normalization factor . The field $\chi(x)$ is constrained to equal (χ_L, χ_R) on the surface $X^0 = 0$. Finally the action I_E is evaluated as an integral over the portion of Minkowski space with $X^0 > 0$.

The boost invariance of the original Minkowski action insures that the Euclidean action has four dimensional rotation invariance. In particular, the invariance under ω-translations becomes invariance under rotations in the Euclidean (Z, X^0) plane. This suggests a new way to carry out the path integral. Let us define the Euclidean angle in the (Z, X^0) plane to be θ. The angle θ is the Euclidean analogue of the Rindler time ω. Now let us divide the region $X^0 > 0$ into infinitesimal angular wedges as shown in Figure 3.3.

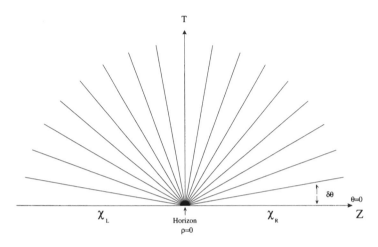

Fig. 3.3   *Euclidean analogue of Rindler space for path integration*

The strategy for computing the path integral is to integrate over the fields in the first wedge between $\theta = 0$ and $\theta = \delta\theta$. The process can be iterated until the entire region $X^0 > 0$ has been covered.

The integral over the first wedge is defined by constaining the fields at $\theta = 0$ and $\theta = \delta\theta$. This defines a *transfer matrix* G in the Hilbert space of the field configuration χ_R. The matrix is recognized to be

$$G = (1 - \delta\theta\, H_R). \tag{3.4.26}$$

To compute the full path integral we raise the matrix G to the power $\frac{\pi}{\delta\omega}$ giving

$$\Psi(\chi_L, \chi_R) = \frac{1}{\sqrt{Z}} \langle \chi_L | e^{-\pi H_R} | \chi_R \rangle \tag{3.4.27}$$

In other words, the path integral defining Ψ is computed as a transition matrix element between initial state χ_R and final state χ_L. The infinitesimal generator which pushes θ surfaces forward is just the Rindler Hamiltonian.

Now we are prepared to compute the density matrix ρ_R. According to the definition in equation 3.3.8, the density matrix ρ_R is given by

$$\rho_R(\chi_R, \chi'_R) = \int \Psi^*(\chi_L, \chi_R)\, \Psi(\chi_L, \chi'_R)\, d\chi_L \qquad (3.4.28)$$

Now using equation 3.4.27 we get

$$\rho_R(\chi_R, \chi'_R) = \tfrac{1}{Z} \int \langle \chi_R | e^{-\pi H_R} | \chi_L \rangle \langle \chi_L | e^{-\pi H_R} | \chi'_R \rangle\, d\chi_L$$

$$= \tfrac{1}{Z} \langle \chi_R | e^{-2\pi H_R} | \chi'_R \rangle \qquad (3.4.29)$$

In other words, the density matrix is given by the operator

$$\rho_R = \frac{1}{Z} exp(-2\pi H_R) \qquad (3.4.30)$$

This remarkable result, discovered by William Unruh in 1976 , says that the Fidos see the vacuum as a *thermal ensemble* with a density matrix of the Maxwell–Boltzmann type. The temperature of the ensemble is

$$T_R = \frac{1}{2\pi} = \frac{1}{\beta_R} \qquad (3.4.31)$$

The derivation of the thermal character of the density matrix and the value of the Rindler temperature in equation 3.4.31 is entirely independent of the particulars of the relativistic field theory. It is equally correct for a free scalar quantum field, quantum electrodynamics, or quantum chromo-dynamics.

3.5   Proper Temperature

It is noteworthy that the temperature T_R is dimensionless. Ordinarily, temperature has units of energy, or equivalently, inverse length. The origin of the dimensionless temperature lies in the dimensionless character of the Rindler time variable ω. Nevertheless we should be able to assign to each Fido a conventional temperature that would be recorded by a standard thermometer held at rest at the location of that Fido. We can consider a thermometer to be a localized object with a set of proper energy levels ϵ_i. The levels ϵ_i are the ordinary energy levels of the thermometer when it is at rest. The thermometer is assumed to be very weakly coupled to the quantum fields so that it eventually will come to thermal equilibrium with them. Let us suppose that the thermometer is at rest with respect to the Fido at position ρ so that it has proper acceleration $\frac{1}{\rho}$. The Rindler energy in equation 3.1.1 evidently receives a contribution from the thermometer of the form

$$H_R(thermometer) = \sum_i \rho \, |i\rangle \, \langle i| \, \epsilon_i \qquad (3.5.32)$$

In other words the Rindler energy level of the ith state of the thermometer is $\rho \, \epsilon_i$.

When the quantum field at Rindler temperature $\frac{1}{2\pi}$ equilibrates with the thermometer, the probability to find the thermometer excited to the ith level is given by the Boltzmann factor

$$P_i = \frac{e^{-2\pi \rho \epsilon_i}}{\sum_j e^{-2\pi \rho \epsilon_j}} \qquad (3.5.33)$$

Accordingly, the thermometer registers a proper temperature

$$T(\rho) = \frac{1}{2\pi \rho} = \frac{1}{\rho} T_R \qquad (3.5.34)$$

Thus each Fido experiences a thermal environment characterized by a temperature which increases as we move toward the horizon at $\rho = 0$. The proper temperature $T(\rho)$ can also be expressed in terms of the proper acceleration of the Fido which is equal to $\frac{1}{\rho}$. Thus, calling the acceleration a, we find

$$T(\rho) = \frac{a(\rho)}{2\pi} \qquad (3.5.35)$$

The reader may wonder about the origin of the thermal fluctuations felt by the Fidos, since the system under investigation is the Minkowski space vacuum. The thermal fluctuations are nothing but the conventional virtual vacuum fluctuations, but now being experienced by accelerated apparatuses. It is helpful in visualizing these effects to describe virtual vacuum fluctuations as short lived particle pairs. In Figure 3.4 ordinary vacuum fluctuations are shown superimposed on a Rindler coordinate mesh. One virtual loop (a) is contained entirely in Region I. That fluctuation can be thought of as a conventional fluctuation described by the quantum Hamiltonian H_R. The fluctuation (b) contained in Region III has no significance to the Fidos in Region I. Finally there are loops like (c) which are partly in Region I but which also enter into Region III. These are the fluctuations which lead to nontrivial entanglements between the degrees of freedom χ_L and χ_R, and which cause the density matrix of Region I to be a mixed state.

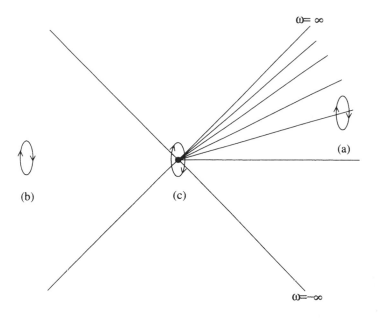

Fig. 3.4 *Vacuum pair fluctuations near the horizon*

A virtual fluctuation is usually considered to be short lived because it "violates energy conservation". If the virtual fluctuation of energy needed to produce the pair is E, then the lifetime of the fluctuation $\sim E^{-1}$.

Now consider the portion of the loop (b) which is found in Region I. From the viewpoint of the Fidos, a particle is injected into the system at $\omega = -\infty$ and $\rho = 0$. The particle travels to some distance and then falls back towards $\rho = 0$ and $\omega = +\infty$. Thus, according to the Fidos, the fluctuation lasts for an infinite time and is therefore not virtual at all. Real particles are seen being injected into the Rindler space from the horizon, and eventually fall back to the horizon. To state it differently, the horizon behaves like a hot membrane radiating and reabsorbing thermal energy.

A natural question to ask is whether the thermal effects are "real". For example, we may ask whether any such thermal effects are seen by freely falling observers carrying their thermometers with them as they pass through the horizon. Obviously the answer is no. A thermometer at rest in an inertial frame in the Minkowski vacuum will record zero temperature. It is tempting to declare that the thermal effects seen by the Fidos are fictitious and that the reality is best described in the frame of the Frefos.

However, by yielding to this temptation we risk prejudicing ourselves too much toward the viewpoint of the Frefos. In particular, we are going to encounter questions of the utmost subtlety concerning the proper relation between events as seen by observers who fall through the horizon of a black hole and those seen by observers who view the formation and evaporation process from a distance. Thus for the moment, it is best to avoid the metaphysical question of whose description is closer to reality. Instead we simply observe that the phenomena are described differently in two different coordinate systems and that different physical effects are experienced by Frefos and Fidos. In particular a Fido equipped with a standard thermometer, particle detector, or other apparatus, will discover all the physical phenomena associated with a local proper temperature $T(\rho) = \frac{1}{2\pi\,\rho}$. By contrast, a Frefo carrying similar apparatuses will see only the zero temperature vacuum state. Later we will discuss the very interesting question of how contradiction is avoided if a Frefo attempts to communicate to the stationary Fidos the information that no thermal effects are present.

We can now state the sense in which a self contained description of Rindler space is possible in ordinary quantum field theory. Since Rindler space has a boundary at $\rho = 0$, a boundary condition of some sort must be provided. We see that the correct condition must be that at some small distance ρ_o, an effective "membrane" is kept at a fixed temperature $T(\rho_o) = \frac{1}{2\pi\,\rho_o}$ by an infinite heat reservoir. It will prove useful later to locate the membrane at a distance of order the Planck length $\ell_P = \sqrt{\hbar G/c^3}$ where quantum gravitational or string effects become important. Such a fictitious membrane at Planckian distance from the horizon is called the *stretched horizon*. We will see later that the stretched horizon has many other physical properties besides temperature, although it is completely unseen by observers who fall freely through it.

Chapter 4

Entropy of the Free Quantum Field in Rindler Space

In the real world, a wide variety of different phenomena take place at different temperature scales. At the lowest temperatures where only massless quanta are produced by thermal fluctuations, one expects to find a very weakly interacting gas of gravitons, photons, and neutrinos. Increase the temperature to the e^+, e^- threshold and electron-positron pairs are produced. The free gas is replaced by a plasma. At higher temperatures, pions are produced which eventually dissociate into quarks and gluons, and so it goes, up the scale of energies. Finally, the Planck temperature is reached where totally new phenomena of an as yet unimagined kind take place.

All of these phenomena have their place in the Fido's description of the region near a horizon. In this lecture we will consider an enormously oversimplified description of the world in which only a single free field is present in a fixed space-time background. There is serious danger in extrapolating far reaching conclusions from so oversimplified a situation. In fact, the paradoxes and contradictions associated with black holes, quantum mechanics, and statistical thermodynamics that these lectures are concerned with are largely a consequence of such unjustified extrapolation. Nevertheless, the study of a free quantum field in Rindler space is a useful starting point.

We consider the field theory defined by equation 2.1.18. Fourier decomposing the field χ in equation 2.1.19 leads to the wave equation in equation 2.1.24.

$$-\frac{\partial^2 \chi_k}{\partial u^2} + \left(k^2 e^{2u}\right) \chi_k = \lambda^2 \chi_k \qquad (4.0.1)$$

In order to quantize the field χ it is necessary to provide a boundary condition when $u \to -\infty$. The simplest method of dealing with this region is to introduce a cutoff at some point $u_o = log\,\epsilon$ at which point the field (or its first derivative) is made to vanish. The parameter ϵ represents the proper distance of the cutoff point to the horizon. Physically we are introducing a perfectly reflecting mirror just outside the horizon at a distance ϵ. Later we will remove the cutoff by allowing $u_o \rightarrow -\infty$.

It is by no means obvious that a reflecting boundary condition very near the horizon is a physically reasonable way to regularize the theory. However it will prove interesting to separate physical quantities into those which are sensitive to ϵ and those which are not. Those things which depend on ϵ are sensitive to the behavior of the physical theory at temperatures of order $\frac{1}{2\pi\epsilon}$ and greater.

Each transverse Fourier mode χ_k can be thought of as a free 1+1 dimensional quantum field confined to a box. One end of the box is at the reflecting boundary at $u = u_o = log\,\epsilon$. The other wall of the box is provided by the repulsive potential

$$V(u) = k^2\,exp(2u)$$

which becomes large when $u > -log\,k$. Thus we may approximate the potential by a second wall at $u = u_1 = -log\,k$. The total length of the box depends on k and ϵ according to

$$L(k) = -log(\epsilon\,k) \qquad (4.0.2)$$

For each value of k the field χ_k can be expanded in mode functions and creation and annihilation operators according to

$$\chi_k(u) = \sum_n \left[a^+(n,k)\,f_{n,k}(u) + a^-(n,k)\,f^*_{n,k}(u) \right] \qquad (4.0.3)$$

where the mode (n,k) has frequency $\lambda(n,k)$. The Rindler Hamiltonian is given by

$$H_R = \int d^2k \sum_n \lambda(n,k)\,a^\dagger(n,k)\,a(n,k)$$
$$= \sum_n \lambda(n,k)\,N(n,k) \qquad (4.0.4)$$

where

$$N(n,k) = a^\dagger(n,k)\,a(n,k) \qquad (4.0.5)$$

Thus far the quantization rules are quite conventional. The new and unusual feature of Rindler quantization, encountered in Chapter 3, is that we do not identifiy the vacuum with the state annihilated by the $a(n, k)$, but rather with the thermal density matrix

$$\rho_R = \prod_{n,k} \rho_R(n, k) \qquad (4.0.6)$$

with

$$\rho_R(n, k) \sim exp\left[-2\pi\lambda(n, k)\, a^\dagger(n, k)\, a(n, k)\right] \qquad (4.0.7)$$

Thus the average occupation number of each mode is

$$\langle N(n, k)\rangle = \frac{1}{exp[2\pi\lambda(n, k)] - 1} \qquad (4.0.8)$$

These particles constitute the *thermal atmosphere*.

The reader might wonder what goes wrong if we choose the state which is annihilated by the a's. Such a state is not at all invariant under translations of the original Minkowski coordinates Z and T. In fact, a careful computation of the expectation value of $T^{\mu\nu}$ in this state reveals a singular behavior at the horizon. Certainly this is not a good candidate to represent the original Minkowski vacuum.

A black hole, on the other hand, is not a translationally invariant system. One might therefore suppose that the evolution of the horizon might lead to the Fock space vacuum with no quanta rather than the thermal state. This however would clearly violate the fourth guiding principle stated in the introduction: To a freely falling observer, the horizon of a black hole should in no way appear special. Moreover, the large back reaction on the gravitational field that would result from the divergent expectation value of $T^{\mu\nu}$ makes it unlikely that this state can exist altogether.

Physical quantities in Rindler space can be divided into those which are sensitive to the cutoff at ϵ and those which are not. As an example of insensitive quantities, the field correlation functions such as

$$\langle \chi(X, Y, u)\, \chi(X', Y', u')\rangle = Tr\, \rho\, \chi(X, Y, u)\, \chi(X', Y', u'), \qquad (4.0.9)$$

are found to have smooth limits as $\epsilon \to 0$, as long as the points (X, Y, u) and (X', Y', u') are kept away from the horizon. Therefore such quantities can be said to *decouple* from the degrees of freedom within a distance ϵ of the horizon. A much more singular quantity which will be of great concern in future lectures is the entropy of the vacuum state.

Since the relevant density matrix has the Maxwell–Boltzmann form, we can use equations 3.3.16 and 3.3.17 to obtain the entropy. Defining

$$Tr \, e^{-\beta H} = Z(\beta) \tag{4.0.10}$$

and using the identity

$$\rho \, log \, \rho = \left. \frac{\partial}{\partial N} \rho^N \right|_{N=1} \tag{4.0.11}$$

we obtain

$$S = \left. -Tr \, \frac{\partial}{\partial N} \, \frac{e^{-N\beta H}}{Z(\beta)^N} \right|_{N=1}$$

$$= +Tr \, \beta H \, \frac{e^{-\beta H}}{Z} + lnZ \tag{4.0.12}$$

$$= \beta \langle H \rangle + lnZ$$

Defining $E = \langle H \rangle$ and $F = -\frac{1}{\beta} logZ$ we find the usual thermodynamic identity

$$S = \beta(E - F) \tag{4.0.13}$$

Another identity follows from using $E = -\frac{\partial logZ}{\partial \beta}$ where we find

$$S = -\beta^2 \, \frac{\partial(logZ/\beta)}{\partial \beta} \tag{4.0.14}$$

The entropy S in equations 4.0.13 and 4.0.14 can be thought of as both entanglement and thermal entropy in the special case of the Rindler space density matrix. This is because the effect of integrating over the fields χ_L in equation 3.4.28 is to produce the thermal density matrix in equation 3.4.30. Thus the computation of the entropy of Rindler space is reduced to ordinary thermodynamic methods. For the present case of free fields the entropy is additive over the modes and can be estimated from the formula for the thermodynamics of a free 1+1 dimensional scalar field.

To compute the total entropy we begin by replacing the infinite transverse X, Y plane by a finite torus with periodic boundary conditions. This has the effect of discretizing the values of k. Thus

$$k_X = \frac{2n_X \pi}{B} \quad , \quad k_Y = \frac{2n_Y \pi}{B} \tag{4.0.15}$$

where B is the size of the torus.

The entropy stored in the field χ_k can be estimated from the entropy density of a 1+1 dimensional massless free boson at temperature T. A standard calculation gives the entropy density $\frac{S}{L}$ to be given by

$$\frac{S}{L} = \frac{\pi}{3} T \qquad (4.0.16)$$

where T is the temperature. Substituting $T = \frac{1}{2\pi}$ and equation 4.0.2 for the length L gives the entropy of χ_k

$$S(k) = \frac{1}{6} |log \, k \, \epsilon| \qquad (4.0.17)$$

To sum over the values of k we use equation 4.0.15 and let $B \to \infty$

$$S_{Total} = \frac{B^2}{24\pi^2} \int d^2k \, |log \, k \, \epsilon| \qquad (4.0.18)$$

In evaluating equation 4.0.18, the integral must be cut off when $k > \frac{1}{\epsilon}$. This is because when $k = \frac{1}{\epsilon}$ the potential is already large at $u = u_o$ so that the entire contribution of χ_k is supressed. We find that S is approximately given by

$$S_{Total} \approx \frac{1}{96\pi^2} \frac{B^2}{\epsilon^2} \qquad (4.0.19)$$

From equation 4.0.19 we see two important features of the entropy of Rindler space. The first is that it is proportional to the transverse area of the horizon, B^2. One might have expected it to diverge as the volume of space, but this is not the case. The entropy is stored in the vicinity of the stretched horizon and therefore grows only like the area. The second feature which should alarm us is that the entropy per unit area diverges like $\frac{1}{\epsilon^2}$. As we shall see, the entropy density of the horizon is a physical quantity whose exact value is known. Nevertheless the divergence in S indicates that its value is sensitive to the ultraviolet physics at very small length scales.

Further insight into the form of the entropy can be gained by recalling that the proper temperature $T(\rho)$ is given by $T(\rho) = \frac{1}{2\pi\rho}$. Furthermore the entropy density of a 3+1 dimensional free scalar field is given by

$$S(T) = V \frac{2}{\pi^2} \zeta(4) k_B \left(\frac{k_B T}{\hbar c}\right)^3 = V \frac{2\pi^2}{45} T^3 \qquad (4.0.20)$$

Now consider the entropy stored in a layer of thickness $\delta\rho$ and area B^2 at a distance ρ from the horizon

$$\delta S(\rho) = \frac{2\pi^2}{45} T^3(\rho) \, \delta\rho \, B^2$$

$$= \frac{2\pi^2}{45} \frac{1}{(2\pi\rho)^3} \, \delta\rho \, B^2$$

(4.0.21)

To find the full entropy we integrate with respect to ρ

$$S = \frac{B^2}{(2\pi)^3} \frac{2\pi^2}{45} \int_\epsilon^\infty \frac{d\rho}{\rho^3}$$

$$= \frac{B^2}{(2\pi)^3} \frac{\frac{2\pi^2}{45}}{2\epsilon^2}$$

(4.0.22)

Now we see that the entropy is mainly found near the horizon because that is where the temperature gets large.

4.1   Black Hole Evaporation

The discovery of a temperature seen by an accelerated fiducial observer adds a new dimension to the equivalence principle. We can expect that identical thermal effects will occur near the horizon of a very massive black hole. However, in the case of a black hole a new phenomenon can take place – evaporation. Unlike the Rindler case, the thermal atmosphere is not absolutely confined by the centrifugal potential in equation 2.0.7. The particles of the thermal atmosphere will gradually leak through the barrier and carry off energy in the form of thermal radiation. A good qualitative understanding of the process can be obtained from the Rindler quantum field theory in equation 2.1.22 by observing two facts:

1) The Rindler time ω is related to the Schwarzschild time t by the equation

$$\omega = \frac{t}{4MG}$$

(4.1.23)

Thus a field quantum with Rindler frequency ν_R is seen by the distant Schwarzschild observer to have a red shifted frequency ν

$$\nu = \frac{\nu_R}{4MG}$$

(4.1.24)

The implication of this fact is that the temperature of the thermal atmosphere is reckoned to be red shifted also. Thus the temperature as seen by the distant observer is

$$T = \frac{1}{2\pi} \times \frac{1}{4MG} = \frac{1}{8\pi MG},$$

(4.1.25)

a form first calculated by Stephen Hawking.

2) The centrifugal barrier which is described in the Rindler theory by the potential $k^2 exp(2u)$ is modified at distances $r \approx 3MG$ as in Figure 2.1. In particular the maximum value that V takes on for angular momentum zero is

$$V_{max}(\ell = 0) = \frac{27}{1024} \frac{1}{M^2 G^2}$$

(4.1.26)

Any s-wave quanta with frequencies of order $(V_{max})^{1/2} = \frac{3\sqrt{3}}{32MG}$ or greater will easily escape the barrier. Since the average energies of massless particles in thermal equilibrium at temperature T is of course of order T, equation 4.1.25 indicates that some of the s-wave particles will easily escape to infinity. Unless the black hole is kept in equilibrium by incoming radiation it will lose energy to its surroundings.

Particles of angular momenta higher than s-waves cannot easily escape because the potential barrier is higher than the thermal scale. The black hole is like a slightly leaky cavity containing thermal radiation. Most quanta in the thermal atmosphere have high angular momenta and reflect off the walls of the cavity. A small fraction of the particles carry very low angular momenta. For these particles, the walls are semi-transparent and the cavity slowly radiates its energy. This is the process first discovered by Hawking and is referred to as Hawking radiation.

The above description of Hawking radiation does not depend in any essential way on the free field approximation. Indeed it only makes sense if there are interactions of sufficient strength to keep the system in equilibrium during the course of the evaporation. In fact, most discussions of Hawking radiation rely in an essential way on the free field approximation, and ultimately lead to absurd results. At the end of the next lecture, we will discuss one such absurdity.

Chapter 5

Thermodynamics of Black Holes

We have seen that a large black hole appears to a distant observer as a body with temperature

$$T = \frac{1}{8\pi MG} \qquad (5.0.1)$$

and energy M. It follows thermodynamically that it must also have an entropy. To find the entropy we use the first law of thermodynamics in the form

$$dE = T\,dS \qquad (5.0.2)$$

where E, the black hole energy, is replaced by M. Using equation 5.0.1

$$dM = \tfrac{1}{8\pi MG}\,dS$$

from which we deduce

$$S = 4\pi M^2 G \qquad (5.0.3)$$

The Schwarzschild radius of the black hole is $2MG$ and the area of the horizon is $4\pi(4M^2G^2)$ so that

$$S_{BH} = \frac{Area}{4G} \qquad (5.0.4)$$

This is the famous Bekenstein–Hawking entropy. It is gratifying that it is proportional to the area of the horizon. This, as we have seen, is where all the infalling matter accumulates according to external observers. We have seen in Chapter 4 equation 4.0.19 that the matter fields in the vicinity of the horizon give rise to an entropy. Presumably this entropy is part of the entropy of the black hole, but unfortunately it is infinite as $\epsilon \to 0$. Evidently something cuts off the modes which are very close to the horizon. To get an

idea of where the cutoff must occur, we can require that the contribution
in equation 4.0.19 not exceed the entropy of the black hole S_{BH}

$$\frac{1}{96\pi^2\,\epsilon^2} \underset{\sim}{<} \frac{1}{4G} \qquad\qquad (5.0.5)$$

or

$$\epsilon \underset{\sim}{>} \frac{\sqrt{G}}{15} \qquad\qquad (5.0.6)$$

In other words, the cutoff must not be much smaller than the Planck length,
where the Planck length is given in terms of Newton's constant as $\ell_P = \sqrt{\frac{\hbar}{c^3}G}$. This is of course not surprising. It is widely believed that the nasty
divergences of quantum gravity will somehow be cut off by some mechanism
when the distance scales become smaller than $\sqrt{G}$.

What is the real meaning of the black hole entropy? According to the
principles stated in the introduction to these lectures, the entropy reflects
the number of microscopically distinct quantum states that are "coarse
grained" into the single macroscopic state that we recognize as a black
hole. The number of such states is of order $exp\, S_{B.H.} = exp\left[4\pi M^2 G\right]$.
Another way to express this is through the level density of the black hole

$$\frac{dN}{dM} \sim exp\left[4\pi M^2 G\right] \qquad\qquad (5.0.7)$$

where dN is the number of distinct quantum states with mass M in the
interval dM.

The entropy of a large black hole is an extensive quantity in the sense
that it is proportional to the horizon area. This suggests that we can
understand the entropy in terms of the local properties of a limiting black
hole of infinite mass and area. The entropy diverges, but the entropy per
unit area is finite. The local geometry of a limiting black hole horizon is of
course Rindler space.

Let us consider the Rindler energy of the horizon. By definition it is
conjugate to the Rindler time ω. Accordingly we write

$$[E_R(M), \omega] = i \qquad\qquad (5.0.8)$$

Here E_R is the Rindler energy which is of course the eigenvalue of the
Rindler Hamiltonian. We assume that for a large black hole the Rindler
energy is a function of the mass of the black hole.

The mass and Schwarzschild time are also conjugate

$$[M, t] = i \qquad (5.0.9)$$

Now use $\omega = \frac{t}{4MG}$ to obtain

$$\left[E_R(M), \tfrac{t}{4MG}\right] = i$$

or

$$[E_R(M), t] = 4MG\,i \qquad (5.0.10)$$

Finally, the conjugate character of M and t allows us to write equation 5.0.10 in the form

$$\frac{\partial E_R}{\partial M} = 4MG \qquad (5.0.11)$$

and

$$E_R = 2M^2 G \qquad (5.0.12)$$

The Rindler energy and the Schwarzschild mass are both just the energy of the black hole. The Schwarzschild mass is the energy as reckoned by observers at infinity using t-clocks, while the Rindler energy is the (dimensionless) energy as defined by observers near the horizon using ω-clocks. It is of interest that the Rindler energy is also extensive. The area density of Rindler energy is

$$\frac{E_R}{A} = \frac{1}{8\pi G} \qquad (5.0.13)$$

The Rindler energy and entropy satisfy the first law of thermodynamics

$$dE_R = \frac{1}{2\pi}\,dS \qquad (5.0.14)$$

where $\frac{1}{2\pi}$ is the Rindler temperature. Thus we see the remarkable fact that horizons have universal local properties that behave as if a thermal membrane or stretched horizon with real physical properties were present. As we have seen, the stretched horizon also radiates like a black body.

The exact rate of evaporation of the black hole is sensitive to many details, but it can easily be estimated. We first recall that only the very low angular momentum quanta can escape the barrier. For simplicity, suppose that only the s-wave quanta get out. The s-wave quanta are described in terms of a 1+1 dimensional quantum field at Rindler temperature $\frac{1}{2\pi}$. In the same units, the barrier height for the s-wave quanta is comparable to the temperature. It follows that approximately one quantum per unit Rindler time will excape. In terms of the Schwarzschild time, the flux of quanta is of order $\frac{1}{MG}$. Furthermore each quantum carries an energy at infinity of order the Schwarzschild temperature $\frac{1}{8\pi MG}$. The resulting rate of energy loss is of order $\frac{1}{M^2G^2}$. We call this L, the *luminosity*. Evidently energy conservation requires the black hole to lose mass at just this rate

$$\frac{dM}{dt} = -L = -\frac{C}{M^2G^2} \tag{5.0.15}$$

where C is a constant of order unity. The constant C depends on details such as the number of species of particles that can be treated as light enough to be thermally produced. It is therefore not really constant. When the mass of the black hole is large and the temperature low, only a few species of massless particles contribute and C is constant.

If we ignore the mass dependence of C, equation 5.0.15 can be integrated to find the time that the black hole survives before evaporating to zero mass. This evaporation time is evidently of order

$$t_{evaporation} \sim M^3 G^2 \tag{5.0.16}$$

It is interesting that luminosity in equation 5.0.15 is essentially the Stephan–Boltzmann law

$$L \sim T^4 \cdot Area \tag{5.0.17}$$

Using $T \sim \frac{1}{MG}$ and $Area \sim M^2G^2$ in equation 5.0.15 gives equation 5.0.17. However the physics is very different from that of a radiating star. In that case the temperature and size of the system are related in an entirely different way. The typical wavelength of a photon radiated from the sun is $\sim 10^{-5}$ cm, while the radius of the surface of the sun is $\sim 10^{11}$ cm. The sun is for all intents and purposes infinite on the scale of the emitted photon wavelengths. The black hole on the other hand emits quanta of wavelength $\sim \frac{1}{T} \sim MG$, which is about equal to the Schwarzschild radius. Observing a black hole by means of its Hawking radiation will always produce a fuzzy image, unlike the image of the sun.

Chapter 6

Charged Black Holes

There are a variety of ways to generalize the conventional Schwarzschild black hole. By going to higher dimensions we can consider not only black holes, but black strings, black membranes, and so forth. Typically black strings and branes are studied as systems of infinite extent, and therefore have infinite entropy. For this reason they can store infinite amounts of information. Higher dimensional Schwarzschild black holes are quite similar to their four-dimensional counterparts.

Another way to generalize the ordinary black hole is to allow it to carry gauge charge and/or angular momentum. In this lecture we will describe the main facts about charged black holes. The most important fact about them is that they cannot evaporate away completely. They have ground states with very special and simplifying features.

Thus, let us consider electrically charged black holes. The metric for a Reissner–Nordstrom black hole is

$$ds^2 = -\left(1 - \frac{2MG}{r} + \frac{Q^2 G}{r^2}\right) dt^2 + \left(1 - \frac{2MG}{r} + \frac{Q^2 G}{r^2}\right)^{-1} dr^2 + r^2 \, d\Omega^2$$

$$(6.0.1)$$

The electric field is given by the familiar Coulomb law

$$\begin{aligned} E_r &= \frac{Q}{r^2} \\ E_{\theta,\phi} &= 0 \end{aligned}$$

$$(6.0.2)$$

If the electric field is too strong at the horizon, it will cause pair production of electrons, which will discharge the black hole in the same manner as a nucleus with $Z >> 137$ is discharged. Generally the horizon occurs at $r \sim MG$, and the threshold field for unsupressed pair production is $E \sim m_e^2$, where m_e is the electron mass. Pair production is exponentially suppressed if

$$\frac{Q}{M^2 G^2} \ll m_e^2 \tag{6.0.3}$$

or alternativley $\frac{M^2}{Q} \gg \frac{1}{m_e^2 G^2}$.

For $Q^2 > M^2 G$ the metric in equation 6.0.1 has a time-like singularity with no horizon to cloak it. Such "naked singularities" indicate a breakdown of classical relativity visible to a distant observer. The question is not whether objects with $Q^2 > M^2 G$ can exist. Clearly they can. The electron is such an object. The question is whether they can be described by classical general relativity. Clearly they cannot. Accordingly we restrict our attention to the case $M^2 > \frac{Q^2}{G}$ or $\frac{M^2}{Q} > \frac{Q}{G}$. A Reissner–Nordstrom black hole that saturates this relationship $M^2 = \frac{Q^2}{G}$ is called an *extremal black hole*. Thus equation 6.0.3 is satisfied if

$$Q \gg \frac{1}{m_e^2 G} \sim 10^{44}$$

Black holes with charge $\gg 10^{44}$ can only discharge by exponentially suppressed tunneling processes. For practical purposes we regard them as stable.

The Reissner–Nordstrom solution has two horizons, an outer one and an inner one. They are defined by

$$\left(1 - \frac{2MG}{r_\pm} + \frac{Q^2 G}{r_\pm^2}\right) = 0 \tag{6.0.4}$$

where $r_+ \, (r_-)$ refers to the outer (inner) horizon:

$$r_\pm = MG\left[1 \pm \sqrt{1 - \frac{Q^2}{M^2 G}}\right] \tag{6.0.5}$$

The metric can be rewritten in the form

$$ds^2 = -\frac{(r - r_+)(r - r_-)}{r^2}\, dt^2 + \frac{r^2\, dr^2}{(r - r_+)(r - r_-)} + r^2\, d\Omega^2 \tag{6.0.6}$$

Note that in the *extremal limit* $M^2 = \frac{Q^2}{G}$ the inner and outer horizons merge at $r_\pm = MG$.

To examine the geometry near the outer horizon, let us begin by computing the distance from r_+ to an arbitrary point $r > r_+$. Using equation 6.0.6

we compute the distance ρ to be

$$\rho = \int \frac{r}{\sqrt{(r-r_+)(r-r_-)}} \, dr \qquad (6.0.7)$$

We define the following

$$\begin{aligned}
r_+ + r_- &\equiv \Sigma \\
r_+ - r_- &\equiv \Delta \\
y &\equiv r - \tfrac{\Sigma}{2}
\end{aligned} \qquad (6.0.8)$$

We find

$$\rho = \int \frac{y+\frac{\Sigma}{2}}{\sqrt{y^2 - \left(\frac{\Delta}{2}\right)^2}} \, dy$$

$$= \sqrt{y^2 - \frac{\Delta^2}{4}} + \frac{\Sigma}{2} \cosh^{-1}\left(\frac{2}{\Delta}y\right) \qquad (6.0.9)$$

The radial-time metric is given by

$$ds^2 = -\frac{\left(y^2 - \frac{\Delta^2}{4}\right)}{\left(y+\frac{\Sigma}{2}\right)^2} \, dt^2 + d\rho^2 \qquad (6.0.10)$$

Expanding equation 6.0.9 near the horizon r_+ one finds

$$\rho \approx \left(y - \frac{\Delta}{2}\right)^{1/2} \left(\frac{2r_+}{\Delta^{1/2}}\right). \qquad (6.0.11)$$

Note that the proper distance becomes infinite for extremal black holes. For non-extremal black holes, equation 6.0.10 becomes

$$ds^2 \cong \frac{\Delta^2}{4r_+^4} \rho^2 \, dt^2 - d\rho^2$$

$$\cong \rho^2 \, d\omega^2 - d\rho^2 \qquad (6.0.12)$$

where

$$\omega \equiv \frac{\Delta}{2r_+^2} t \qquad (6.0.13)$$

Evidently the horizon geometry is again well approximated by Rindler space. The charge density on the horizon is $\frac{Q}{4\pi r_\perp^2}$. Since $r_+ \sim MG$ the charge density is $\sim \frac{Q}{4\pi M^2 G^2}$. Thus for near extremal black holes, the charge density is of the form

$$\frac{Q}{4\pi r_+^2} \sim \frac{M\sqrt{G}}{4\pi M^2 G^2} \cong \frac{1}{4\pi M G^{3/2}} \qquad (6.0.14)$$

For very massive black holes the charge density becomes vanishingly small. Therefore the local properties of the horizon cannot be distinguished from those of a Schwarzschild black hole. In particular, the temperature at a small distance ρ_o from the horizon is $\frac{1}{2\pi\rho_o}$. From equation 6.0.13 we can compute the temperature as seen at infinity.

$$T(\infty) = \frac{\Delta}{2r_+^2}\frac{1}{2\pi} \qquad (6.0.15)$$

Using

$$\Delta = 2MG\sqrt{1 - \frac{Q^2}{M^2 G}}$$

$$r_+ = MG\left[1 + \sqrt{1 - \frac{Q^2}{M^2 G}}\right] \qquad (6.0.16)$$

We find

$$T(\infty) = \frac{2MG\sqrt{1 - \frac{Q^2}{M^2 G}}}{4\pi M^2 G^2 \left[1 + \sqrt{1 - \frac{Q^2}{M^2 G}}\right]^2} \qquad (6.0.17)$$

As the black hole tends to extremality, the horizon becomes progressively more removed from any fiducial observer. From equation 6.0.9 we see that as $\Delta \to 0$

$$\rho \to y + \frac{\Sigma}{2}log(2y) - log\Delta \qquad (6.0.18)$$

Thus for a fiducial observer at a fixed value of r the horizon recedes to infinite proper distance as $\Delta \to 0$.

In the limit $\Delta \to 0$ the geometry near the horizon simplifies to the form

$$ds^2 = \left[-\left(r_+ sinh\frac{\rho}{r_+}\right)^2 d\omega^2 + d\rho^2\right] + r_+^2 d\Omega^2 \qquad (6.0.19)$$

which, although infinitely far from any fiducial observer with $r \neq r_+$, is approximately Rindler.

We note from equation 6.0.17 that in the extremal limit the temperature at infinity tends to zero. The entropy, however, does not tend to zero. This can be seen in two ways, by focusing either on the region very near the horizon or the region at infinity. As we have seen, the local properties of the horizon even in the extreme limit are identical to the Schwarzschild case from which we deduce an entropy density $\frac{1}{4G}$. Accordingly,

$$S = \frac{Area}{4G} = \frac{\pi r_+^2}{G} = \pi M^2 G \qquad (6.0.20)$$

We can deduce this result by using the first law together with equation 6.0.17

$$dM = T\,dS \qquad \text{(Fixed Q)}$$

to obtain $S = \frac{Area}{4G}$ as a general rule.

The fact that the temperature goes to zero in the extreme limit indicates that the evaporation process slows down and does not proceed past the point $Q = M\sqrt{G}$. In other words, the extreme limit can be viewed as the ground state of the charged black hole. However it is unusual in that the entropy does not also tend to zero. This indicates that the ground state is highly degenerate with a degeneracy $\sim e^S$. Whether this degeneracy is exact or only approximate can not presently be answered in the general case. However in certain supersymmetric cases the supersymmetry requires exact degeneracy.

The metric in equation 6.0.19 for extremal black holes can be written in a form analogous to equations 1.3.18 and 1.4.21 by introducing a radial variable

$$\frac{R}{r_+} = \frac{e^{\rho/r_+} - 1}{e^{\rho/r_+} + 1} \qquad (6.0.21)$$

The metric then takes the form

$$d\tau^2 = \left[\left(\frac{2}{1 - \frac{R^2}{r_+^2}} \right) (R^2\, d\omega^2 - dR^2) \right] - r_+^2\, d\Omega^2 \qquad (6.0.22)$$

Obviously the physics near $\frac{R}{r_+} \to 0$ is identical to Rindler space, from which it follows that the horizon will have the usual properties of temperature, entropy, and a thermal atmosphere including particles of high angular momenta trapped near the horizon by a centrifugal barrier.

Although the external geometry of an extreme or near extreme Reissner–Nordstrom black hole is very smooth with no large curvature, one can nevertheless expect important quantum effects in its structure. To understand why, consider the fact that as $\Delta \to 0$ the horizon recedes to infinity. Classically, if we drop the smallest amount of energy into the extreme black hole, the location of the horizon, as measured by its proper distance, jumps an infinite amount. In other words, the location of the horizon of an extremal black hole is very unstable. Under these circumstances, quantum fluctuations can be expected to make the location very uncertain. Whether this effect leads to a lifting of the enormous degeneracy of ground states or any other physical phenomena is not known at present except in the supersymmetric case.

Chapter 7

The Stretched Horizon

Thus far our description of the near-horizon region of black holes, or Rindler space, has been in terms of quantum field theory in a fixed background geometry. But we have already run into a contradiction in applying quantum field theory, although we didn't spell it out. The problem arose in Chapter 4 when we found that the entropy per unit area of the horizon diverges as the cutoff ϵ tends to zero (see equation 4.0.22). That in itself is not a problem. What makes it a problem is that we later found that black hole thermodynamics requires the entropy to be $\frac{A}{4G\hbar}$. Free quantum field theory is giving too much entropy in modes very close to the horizon, where the local temperature diverges. The fact that the entropy is infinite in quantum field theory implies that any quantity that depends on the finiteness of the entropy will be miscalculated using quantum field theory.

One possibility is that we have overestimated the entropy by assuming free field theory. Equation 4.0.20 could be modified by interactions. Indeed that is so, but the effect goes in the wrong direction. The correct entropy density for a general field theory can always be parameterized by

$$S(T) = \gamma(T) T^3$$

where $\gamma(T)$ represents the number of "effective" degrees of freedom at temperature T. It is widely accepted and in many cases proven that $\gamma(T)$ is a monotonically increasing function of T. Thus, conventional interactions are only likely to make things worse. What we need is some new kind of theory that has the effective number of degrees of freedom going to zero very close to the horizon. Let's suppose that ordinary quantum field theory is adequate down to distance scale ϵ. In order that the entropy at distance greater than ϵ not exceed the Bekenstein–Hawking value, we must have the rough inequality

$$\epsilon^2 \stackrel{<}{\sim} G\hbar = \ell_P^2$$

Evidently at distances less than $\sqrt{G\hbar}$ from the horizon the degrees of freedom must be very sparse, or even nonexistent. This leads to the idea that the mathematical horizon should be replaced by an effective membrane, or "stretched" horizon at a distance of roughly one Planck length from the mathematical horizon.

Stretching the horizon has another benefit. Instead of being light-like, a system at the stretched horizon is time-like. This means that real dynamics and evolution can take place on the stretched horizon. As we will see, the stretched horizon has dynamics of its own that includes such phenomena as viscosity and electrical conductivity.

To see that the horizon of a black hole has electrical properties, it is sufficient to study electrodynamics in Rindler space. First let us define the stretched horizon. The metric is

$$d\tau^2 = \rho^2 \, d\omega^2 - d\rho^2 - dx_\perp^2 \qquad (7.0.1)$$

The stretched horizon is just the surface

$$\rho = \rho_o \qquad (7.0.2)$$

where ρ_o is a length of order the Planck length.

The action for the electromagnetic field in Rindler space is

$$W = \int \left[\frac{\sqrt{-g}}{16\pi} g^{\mu\nu} g^{\sigma\tau} F_{\mu\sigma} F_{\nu\tau} + j^\mu A_\mu \right] d\omega \, d\rho \, d^2 x_\perp \qquad (7.0.3)$$

or, substituting the form of the metric

$$W = \int \left[\frac{1}{8\pi} \left(\frac{\left(\dot{\vec{A}} + \vec{\nabla}\phi \right)^2}{\rho} - \rho \left(\vec{\nabla} \times \vec{A} \right)^2 \right) + j \cdot A \right] d\omega \, d\rho \, d^2 x_\perp$$

$$(7.0.4)$$

where $\dot{\vec{A}}$ means $\frac{\partial \vec{A}}{\partial \omega}$ and $\phi = -A_0$, and j is a conserved current in the usual sense $\partial_\mu j^\mu = 0$. As usual

$$\vec{E} = -\vec{\nabla}\phi - \dot{\vec{A}}$$
$$\vec{B} = \vec{\nabla} \times \vec{A}$$

With these definitions, the action becomes

$$W = \int \left[\frac{1}{8\pi} \left(\frac{\left|\vec{E}\right|^2}{\rho} - \rho \left|\vec{B}\right|^2 \right) + j \cdot A \right] d\omega \, d\rho \, d^2 x_\perp \qquad (7.0.5)$$

and the Maxwell equations are

$$\frac{1}{\rho} \dot{\vec{E}} - \vec{\nabla} \times (\rho B) = -4\pi j$$

$$\dot{B} + \vec{\nabla} \times \vec{E} = 0$$

$$\vec{\nabla} \cdot \left(\frac{1}{\rho} \vec{E} \right) = 4\pi j^0 \qquad (7.0.6)$$

$$\vec{\nabla} \cdot \vec{B} = 0$$

We begin by considering electrostatics. By electrostatics we mean the study of fields due to stationary or slowly moving charges placed outside the horizon. Since the charges are slowly moving in Rindler coordinates, it means that they are experiencing proper acceleration. We also assume all length scales associated with the charges are much larger than ρ_o. In particular, the distance of the charges from the stretched horizon is macroscopic.

The surface charge density on the stretched horizon is easily defined. It is just the component of the electric field perpendicular to the stretched horizon, or more precisely

$$\sigma = \left. \frac{1}{4\pi\rho} E_\rho \right|_{\rho=\rho_o}$$

$$= \left. - \frac{1}{4\pi\rho} \partial_\rho \phi \right|_{\rho=\rho_o} \qquad (7.0.7)$$

Working in the Coulomb gauge, the third expression in equation 7.0.6 becomes

$$\vec{\nabla} \cdot \left(\frac{1}{\rho} \vec{E} \right) = -\vec{\nabla} \cdot \left(\frac{1}{\rho} \vec{\nabla} \phi \right) = 0 \qquad (7.0.8)$$

near the stretched horizon. Thus

$$\partial_\rho^2 \phi - \frac{1}{\rho} \partial_\rho \phi = -\nabla_\perp^2 \phi \qquad (7.0.9)$$

We can solve this equation near the horizon by the ansatz $\phi \sim \rho^\alpha$. The right hand side will be smaller than the left hand side by 2 powers of ρ and can therefore be ignored. We easily find that $\alpha = 0$ or $\alpha = 2$. Thus we assume

$$\phi = F(x_\perp) + \rho^2\, G(x_\perp) + terms\, higher\, order\, in\, \rho \qquad (7.0.10)$$

Plugging equation 7.0.10 into equation 7.0.9 and evaluating at $\rho = \rho_o$ gives

$$\nabla_\perp^2 F + \rho_o^2\, \nabla_\perp^2 G = 0 \qquad (7.0.11)$$

If ρ_o is much smaller than all other length scales, then equation 7.0.11 is simplified to

$$\nabla_\perp^2 F = 0 \qquad (7.0.12)$$

A similar equation can also be derived for the finite mass black hole.

Since the black hole horizon is compact, equation 7.0.12 proves that $\phi = constant$ on the horizon. This is an interesting result, which proves that the horizon behaves like an electrical conductor.

We can easily identify the surface current density. Taking the time derivative of equation 7.0.7 and using Maxwell's equations 7.0.6 gives

$$4\pi\, \dot\sigma = \frac{1}{\rho_o}\, \dot E_\rho = \left(\vec\nabla \times \rho\vec B\right)_\rho \qquad (7.0.13)$$

Evidently this is a continuity equation if we define:

$$4\pi\, j_x = -\rho\, B_y$$

$$\qquad (7.0.14)$$

$$4\pi\, j_y = \rho\, B_x$$

Now let us consider an electromagnetic wave propagating toward the stretched horizon along the ρ axis. From Maxwell's equations we obtain

$$\dot B_x = \partial_\rho\, E_y$$

$$\dot B_y = -\partial_\rho\, E_x$$

$$\qquad (7.0.15)$$

$$\frac{1}{\rho}\dot E_x = -\partial_\rho\, (\rho\, B_y)$$

$$\frac{1}{\rho}\dot E_y = \partial_\rho\, (\rho\, B_x)$$

To make the equation more familiar, we can redefine the magnetic field

$$\rho \vec{B} = \vec{\beta} \tag{7.0.16}$$

and use tortoise coordinates

$$u = \log \rho$$

Equation 7.0.15 then becomes

$$\dot{\beta}_x = \partial_u E_y$$

$$\dot{\beta}_y = -\partial_u E_x$$

$$\dot{E}_x = \partial_u \beta_y \tag{7.0.17}$$

$$\dot{E}_y = -\partial_u \beta_x$$

The mathematical equations allow solutions in which the wave propagates in either direction along the u-axis. However the physics only makes sense for waves propagating toward the horizon from outside the black hole. For such waves, the Maxwell equations 7.0.17 give

$$\beta_x = E_y$$

$$\beta_y = -E_x \tag{7.0.18}$$

or from equation 7.0.14

$$j_x = \frac{1}{4\pi} E_x$$

$$j_y = \frac{1}{4\pi} E_y$$

Evidently the horizon is an ohmic conductor with a resistivity of 4π. That corresponds to a surface resistance of $377\Omega/square$.* For example, if a circuit is constructed as in Figure 7.1, a current will flow precisely as if the horizon were a conducting surface.

As a last example let us consider dropping a charged point particle into the horizon. Since the horizon is just ordinary flat space, one might conclude that the point charge just asymptotically approaches the horizon

---

*The unit $\Omega/square$ is not a misprint. The resistance of a two-dimensional resistor is scale invariant and only depends on the shape.

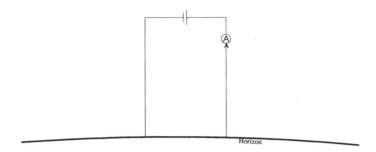

Fig. 7.1   *Battery, ammeter attached to horizon*

with the transverse charge density remaining point-like. However this is not at all what happens on the stretched horizon. This process is shown in Figure 7.2. Without loss of generality, we can take the charge to be at rest at position z_o in Minkowski coordinates. To compute the surface charge density on the stretched horizon, we need to determine the field component E_ρ. The calculation is easy because at any given time the Rindler coordinates are related to the Minkowski coordinates by a boost along the z-axis. Since the component of electric field along the boost direction is invariant, we can write the standard Coulomb field

$$E_\rho = E_z$$

$$= \frac{e\,(z-z_o)}{\left[(z-z_o)^2 + x_\perp^2\right]^{3/2}} \tag{7.0.19}$$

$$= \frac{e\,(\rho\cosh\omega - z_o)}{\left[(\rho\cosh\omega - z_o)^2 + x_\perp^2\right]^{3/2}}$$

Using $4\pi\sigma = \left.\frac{E_\rho}{\rho}\right|_{\rho_o}$ we find

$$\sigma = \frac{e}{4\pi\rho_o} \frac{\rho_o\cosh\omega - z_o}{\left[(\rho_o\cosh\omega - z_o)^2 + x_\perp^2\right]^{3/2}} \tag{7.0.20}$$

Now let's consider the surface density for large Rindler time.

$$\sigma = \frac{e}{4\pi\rho_o} \frac{\rho_o\,e^\omega}{[\rho_o^2\,e^{2\omega} + x_\perp^2]^{3/2}} \tag{7.0.21}$$

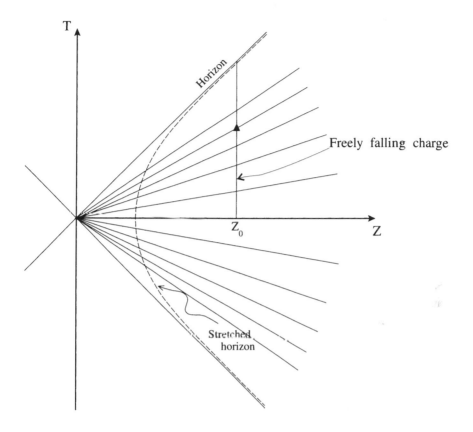

Fig. 7.2 *Charge falling past the stretched horizon*

To better understand this expression, it is convenient to rescale $x_\perp$ using $x_\perp = e^\omega y_\perp$ to obtain

$$\sigma = \frac{e}{4\pi} \frac{e^{-2\omega}}{\left(\rho_o^2 + y_\perp^2\right)^{3/2}} \qquad (7.0.22)$$

It is evident that the charge spreads out at an exponential rate with Rindler time. For a real black hole, it would spread over the horizon in a time

$$\omega = log\,(R_s - \rho_o) = log\,(2MG - \rho_o)$$

or in terms of the Schwarzschild time

$$t = 4MG\omega = 4MG\,log\,(2MG - \rho_o)$$

The exponential spreading of the charge is characteristic of an Ohm's law conductor. To see this we use Ohm's law $j = conductivity\, E$. Taking the divergence gives $\vec{\nabla} \cdot \vec{j} \sim \vec{\nabla} \cdot \vec{E} \sim \sigma$. Now use the continuity equation get the relation $\dot{\sigma} \sim -\sigma$. Evidently the surface charge density will exponentially decrease, and conservation of charge will cause it to spread exponentially.

Thus we see that the horizon has the properties of a more or less conventional hot conducting membrane. In addition to temperature, entropy, and energy, it exhibits dissipative effects such as electrical resistivity and viscosity. The surprising and puzzling thing is that they are completely unnoticed by a freely falling observer who falls through the horizon!

Chapter 8

The Laws of Nature

In this chapter, we want to review three fundamental laws of nature whose compatibility has been challenged. These laws are:

1)   The principle of information conservation
2)   The equivalence principle
3)   The quantum xerox principle

8.1   Information Conservation

In both classical and quantum mechanics there is a very precise sense in which information is never lost from a closed isolated system. In classical physics the principle is embodied in Liouville's theorem: the conservation of phase space volume. If we begin following a system with some limited knowledge of its exact state, we might represent this by specifying an initial region $\Gamma(0)$ in the system's phase space. The region $\Gamma(0)$ has a volume V_{Γ} in the phase space.

Now we let the system evolve. The region $\Gamma(0) = \Gamma$ evolves into the region $\Gamma(t)$. Liouville's theorem tells us that the volume of $\Gamma(t)$ is exactly the same as that of Γ. In this sense the amount of information is conserved.

In a practical sense, information is lost because for most cases of interest the region Γ becomes very complicated like a fractal, and if we coarse grain the phase space, it will appear that Γ is growing. As a definition of *coarse graining*, if one takes every point in the phase space and surrounds it by solid spheres of fixed volume, the union of those spheres is the "coarse grained" volume of phase space, which indeed grows. This is the origin of the second law of thermodynamics. This is illustrated in Figure 8.1.

In quantum mechanics, the conservation of information is expressed as

Fig. 8.1   *Evolution of a fixed volume in phase space*

the unitarity of the S-matrix. If we again approach a system with limited knowledge, we might express this by a projection operator onto a subspace, P, instead of a definite state. The analog of the phase space volume is the dimensionality of the subspace

$$N = Tr\, P$$

The unitarity of the time evolution insures that N is conserved with time.

A more refined definition of information is provided by the concept of entropy. Suppose that instead of specifying a region Γ in phase space, we instead specify a probability density $\rho(p,\, q)$ in phase space. A generalization of the volume is given by the exponential of the entropy $V_\Gamma \to exp\, S$, where

$$S = -\int dp\, dq\, \rho(p,\, q)\, log\, \rho(p,\, q) \qquad (8.1.1)$$

It is easy to check that if $\rho = \frac{1}{V_\Gamma}$ inside Γ and zero outside, then $S = log\, V_\Gamma$.

Similarly, for quantum mechanics the sharp projector P can be replaced by a density matrix ρ. In this case the fine grained or Von Neumann entropy is

$$S = -Tr\, \rho\, log\, \rho \qquad (8.1.2)$$

For the case $\rho = \frac{P}{Tr\, P}$ the entropy is $log\, N$. Thus the entropy is an estimate of the logarithm of the number of quantum states that make up the initial ensemble. In both quantum mechanics and classical mechanics the equations of motion insure the exact conservation of S for a closed, isolated system.

8.2 Entanglement Entropy

In classical physics, the only reason for introducing a phase space probability is a lack of detailed knowledge of the state. In quantum mechanics, there is another reason, entanglement. *Entanglement* refers to quantum correlations between the system under investigation and a second system. More precisely, it involves separating a system into two or more subsystems.

Consider a composite system composed of 2 subsystems A and B. The subsystem $A(B)$ is described by some complete set of commuting observables $\alpha\,(\beta)$. Let us assume that the composite system is in a pure state with wave function $\Psi(\alpha,\beta)$. Consider now the subsystems separately. All measurements performed on $A(B)$ are describable in terms of a density matrix $\rho_A\,(\rho_B)$.

$$(\rho_A)_{\alpha\alpha'} = \sum_\beta \Psi^*(\alpha,\beta)\,\Psi(\alpha',\beta)$$

$$(\rho_B)_{\beta\beta'} = \sum_\alpha \Psi^*(\alpha,\beta)\,\Psi(\alpha,\beta')$$

(8.2.3)

The fact that a subsystem is described by a density matrix and not a pure state may not be due to any lack of knowledge of the state of the composite system. Even in the case of a pure state, the constituent subsystems are generally not described by pure states. The result is an "entanglement entropy" for the subsystems.

Let us consider some properties of the density matrix. For definiteness, consider ρ_A, but we could equally well focus on ρ_B.

1) The density matrix is Hermitian

$$(\rho_A)_{\alpha\alpha'} = (\rho_A)^*_{\alpha'\alpha}$$

(8.2.4)

2) The density matrix is positive semidefinite. This means its eigenvalues are all either positive or zero.
3) The density matrix is normalized to 1.

$$Tr\,\rho = 1$$

(8.2.5)

It follows that all the eigenvalues are between zero and one. If one of the eigenvalues of ρ_A is equal to 1, all the others must vanish. In this case the subsystem A is in a pure state. This only happens if the composite wave function factorizes

$$\Psi(\alpha,\beta) = \psi_A(\alpha)\,\psi_B(\beta)$$

(8.2.6)

In this case B is also in a pure state.

4) The nonzero eigenvalues of ρ_A and ρ_B are equal if the composite system is in a pure state. To prove this, we start with the eigenvalue equation for ρ_A. Call ϕ the eigenvector of ρ_A. Then the eigenvalue condition is

$$\sum_{\beta\alpha'} \Psi^*(\alpha,\beta)\, \Psi(\alpha',\beta)\, \phi(\alpha') = \lambda\, \phi(\alpha)$$

We assume $\lambda \neq 0$. Now we define a candidate eigenvector of ρ_B by

$$\chi(\beta') \equiv \sum_{\alpha'} \Psi^*(\alpha',\beta')\, \phi^*(\alpha').$$

Then

$$\sum_{\beta'} (\rho_B)_{\beta\beta'}\, \chi(\beta') = \sum_{\alpha\beta'} \Psi^*(\alpha,\beta)\, \Psi(\alpha,\beta')\, \chi(\beta')$$

$$= \sum_{\alpha\alpha'\beta'} \Psi^*(\alpha,\beta)\, \Psi(\alpha,\beta')\, \Psi^*(\alpha',\beta')\, \phi^*(\alpha')$$

$$= \lambda \sum_\alpha \Psi^*(\alpha,\beta)\, \phi^*(\alpha)$$

$$= \lambda\, \chi(\beta)$$

Thus $\chi(\beta)$ is an eigenvector of ρ_B with eigenvalue λ.

From the equality of the non-vanishing eigenvalues of ρ_A and ρ_B an important property of entanglement entropy follows:

$$S_A = -Tr\, \rho_A\, log\, \rho_A = S_B \tag{8.2.7}$$

Thus we can just refer to the entanglement entropy as S_E.

The equality of S_A and S_B is only true if the combined state is pure. In that case, the entropy of the composite system vanishes

$$S_{A+B} = 0$$

Evidently entropy is not additive in general.

Next, let us consider a large system Σ that is composed of many similar small subsystems σ_i. Let us suppose the subsystems weakly interact, and the entire system is in a pure state with total energy E. Each subsystem on the average will have energy ϵ.

It is a general property of most complex interacting systems that the density matrix of a small subsystem will be thermal

$$\rho_i = \frac{e^{-\beta\, H_i}}{Z_i} \tag{8.2.8}$$

where H_i is the energy of the subsystem. The thermal density matrix maximizes the entropy for a given average energy ϵ. In general large subsystems or the entire system will not be thermal. In fact, we will assume that the entire system Σ is in a pure state with vanishing entropy.

The *coarse grained* or *thermal entropy* of the composite system is defined to be the sum of the entropies of the small subsystems

$$S_{Thermal} = \sum_i S_i \qquad (8.2.9)$$

By definition it is additive. The coarse grained entropy is what we usually think of in the context of thermodynamics. It is not conserved. To see why, suppose we start with the subsystems in a product state with no corelations. The entropy of each subsystem S_i as well as the entropy of the whole system Σ given by σ_Σ, and the coarse grained entropy of Σ all vanish.

Now the subsystems interact. The wave function develops correlations, meaning that it now fails to factorize. In this case, the subsystem entropies become nonzero

$$S_i \neq 0$$

and the coarse grained entropy also becomes nonzero

$$S_{Thermal} = \sum_i S_i \neq 0$$

However, the "fine grained" entropy of Σ is exactly conserved and therefore remains zero.

Let us consider an arbitrary subsystem Σ_1 of Σ which may consist of one, many, or all of the subsystems σ_i. Typically the fine grained entropy of Σ_1 is defined as the entanglement entropy $S(\Sigma_1)$ of Σ_1 with the remaining subsystem $\Sigma - \Sigma_1$. This will always be less than the coarse grained entropy of Σ_1

$$S_{Thermal}(\Sigma_1) > S(\Sigma_1) \qquad (8.2.10)$$

For example, as Σ_1 approaches Σ, the fine grained entropy $S(\Sigma_1)$ will tend to zero.

Another concept that we can now make precise is the information in a subsystem. The information can be defined by

$$I = S_{Thermal} - S \qquad (8.2.11)$$

Often the coarse grained entropy is the thermal entropy of the system, so that the information is the difference between coarse grained and fine grained entropy.

Since typical small subsystems have a thermal density matrix, the information in a small subsystem vanishes. At the opposite extreme the information of the combined system Σ is just its total thermal entropy. It can be thought of as the hidden subtle correlations between subsystems that make the state of Σ pure.

How much information are in a moderately sized subsystem? One might think that the information smoothly varies from zero (for the σ_i) to $S_{Coarse\,Grained}$ (for Σ). However, this is not so. What actually happens is that for subsystems smaller than about $1/2$ of the total system, the information is negligible.

Entropy and information are naturally measured in "bits". A *bit* is the entropy of a two state system if nothing is known[2]. The numerical value of a bit is *log* 2. Typically for subsystems less than half the size of Σ the information is smaller than 1 bit. The subsystem $\frac{1}{2}\Sigma$ has about 1 bit of information. Thus for $\Sigma_1 < \frac{1}{2}\Sigma$

$$S(\Sigma_1) \cong S_{Thermal}(\Sigma_1)$$

$$I(\Sigma_1) \approx 0$$

Next consider a subsystem with $\Sigma_1 > \frac{1}{2}\Sigma$. How much information does it have? To compute it, we use two facts:

$$S(\Sigma - \Sigma_1) = S(\Sigma_1)$$

$$S(\Sigma - \Sigma_1) \cong S_{Thermal}(\Sigma - \Sigma_1)$$

(8.2.12)

Thus

$$S(\Sigma_1) \cong S_{Thermal}(\Sigma - \Sigma_1) \qquad (8.2.13)$$

The coarse grained entropy of $\Sigma - \Sigma_1$ will be of order $(1 - f)\,S_{Thermal}(\Sigma)$, where f is the fraction of the total degrees of freedom contained in Σ_1. Thus, for $\Sigma_1 < \frac{1}{2}\Sigma$ the information in Σ_1 is essentially zero. However for

$\Sigma_1 > \frac{1}{2}\Sigma$ we get the information to be

$$I(\Sigma_1) = S_{Thermal}(\Sigma_1) - S(\Sigma_1)$$

$$= f\, S_{thermal}(\Sigma) - (1 - f)\, S_{Thermal}(\Sigma) \qquad (8.2.14)$$

$$= (2f - 1)\, S_{Thermal}(\Sigma)$$

We will eventually be interested in the information emitted by a black hole when it evaporates. For now let's consider a conventional system which is described by known laws of physics. Consider a box with perfectly reflecting walls. Inside the box we have a bomb which can explode and fill the box with radiation. The box has a small hole that allows the thermal radiation to slowly leak out. The entire system Σ consists of the subsystem B that includes everything in the box. The subsystem A consists of everyting outside of the box, in this case, outgoing photons.

Initially the bomb is in its ground state, and B has vanishing entropy. When the bomb explodes, it fills the box with thermal radiation. The thermal, or coarse grained, entropy of the box increases, but its fine grained entropy does not. Furthermore, no photons have yet escaped, so $S(A) = 0$ at this time.

$$S_{Thermal}(B) \neq 0$$

$$S(B) = 0 \qquad (8.2.15)$$

$$S(A) = 0$$

Next, photons slowly leak out. The result is that the interior and exterior of the box become entangled. The entanglement entropy, which is equal for A and B, begins to increase. The thermal entropy in the box decreases:

$$S_{Entanglement} \neq 0$$

$$S_{Thermal}(B) \neq 0 \qquad (8.2.16)$$

$$S_{Thermal}(A) \neq 0$$

Eventually, all of the photons escape the box. The thermal or coarse grained entropy as well as the fine grained entropy in the box tends to zero. The box is in a pure state; its ground state.

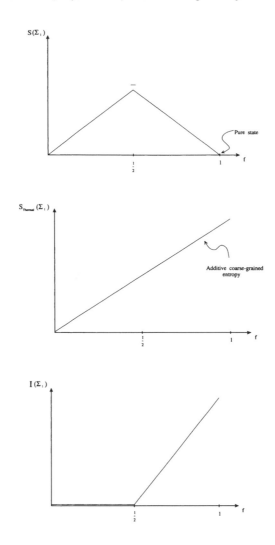

Fig. 8.2   *Top: Von Neumann entropy of Σ_1 vs fraction f. Middle: coarse grained entropy of Σ_1 vs fraction f. Bottom: information vs fraction of total degrees of freedom in Σ_1*

At this time, the thermal or coarse grained entropy of the exterior radiation has increased to its final value. The second law of thermodynamics insures that $S_{Thermal}(A)$ is larger than $S_{Thermal}(B)$ just after the explosion. But the fine grained entropy of A must vanish, since the entanglement has gone to zero.

The actual entanglement entropy must be less than the thermal entropy of A or B. Thus a plot of the various entropies looks like Figure 8.3. Note that the point at which $S_{Thermal}(A) = S_{Thermal}(B)$ defines the time at which the information in the outside radiation begins to grow. Before that point, a good deal of energy has escaped, but no information. Roughly the point where information appears outside of the box is the point where half of the final entropy of the photons has emerged.

It is useful to define this time at which information begins to emerge. This time is called the *information retention time*. It is the amount of time that it takes to retrieve a single bit of information about the initial state of the box.

Thus we see how information conservation works for a conventional quantum system. The consequence of this principle is the final radiation field outside the box must be in a pure state. However, this does not mean that localized regions containing a small fraction of the photons cannot be extremely thermal. They typically carry negligible information.

The description of the evolution of the various kinds of entropy follow from very general principles. Thus we regard the conservation of information in black hole evaporation as a fundamental law of nature. Note that it applies to observations made from outside the black hole.

8.3   Equivalence Principle

A second law of nature concerns the nature of gravitation. In its simplest form the equivalence principle says that a gravitational field is locally equivalent to an accelerated frame. More exactly, it says that a freely falling observer or system will not experience the effects of gravity except through the tidal forces, or equivalently, the curvature components. We have seen that the magnitude of the curvature components at the horizon are small and tend to zero as the mass and radius of the black hole increase. The curvature typically satisfies

$$R \sim \frac{1}{(MG)^2}$$

Any freely falling system of size much smaller than MG will not be distorted or otherwise disrupted by the presence of the horizon.

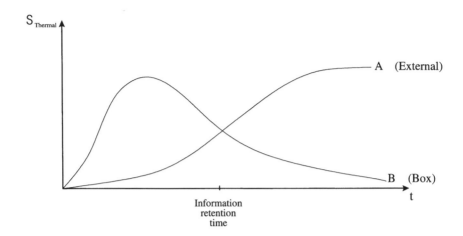

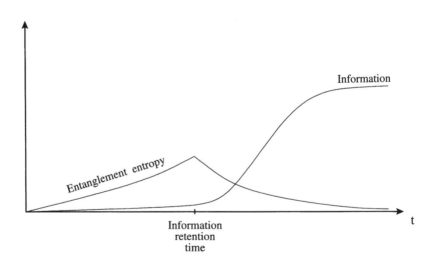

Fig. 8.3 *Top: evolution of the thermal entropies of box and exterior. Bottom: evolution of entanglement entropy and information*

The equivalence principle requires the horizon of a very large black hole to have the same effects on a freely falling observer as the horizon of Rindler space has; namely, no effect at all.

8.4 Quantum Xerox Principle

The third law of nature that plays an important role in the next lecture is the impossibility of faithfully duplicating quantum information. What it says is that a particular kind of apparatus cannot be built. We call it a *Quantum Xerox Machine.** It is a machine into which any system can be inserted, and which will copy that system, producing the original and a duplicate. To see why such a system is not possible, imagine that we insert a spin in the input port as in Figure 8.4. If the spin is in the up state with

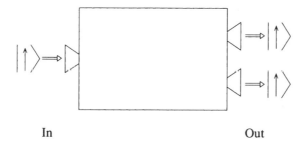

In   Out

Fig. 8.4 *Schematic diagram of quantum Xerox machine*

respect to the z-axis, it is duplicated

$$|\uparrow\rangle \rightarrow |\uparrow\rangle |\uparrow\rangle \tag{8.4.17}$$

Similarly, if it is in the down state, it is duplicated

$$|\downarrow\rangle \rightarrow |\downarrow\rangle |\downarrow\rangle \tag{8.4.18}$$

Now suppose that the spin is inserted with its polarization along the x-axis, i.e.

$$\frac{1}{\sqrt{2}} (|\uparrow\rangle + |\downarrow\rangle) \tag{8.4.19}$$

---

*The quantum Xerox principle is sometimes called the *no-cloning principle*

The general principles of quantum mechanics require the state to evolve linearly. Thus from equations 8.4.17 and 8.4.18

$$\frac{1}{\sqrt{2}} \left(|\uparrow\rangle + |\downarrow\rangle \right) \rightarrow \frac{1}{\sqrt{2}} \left(|\uparrow\rangle \, |\uparrow\rangle + |\downarrow\rangle \, |\downarrow\rangle \right) \tag{8.4.20}$$

On the other hand, a true quantum Xerox machine is required to duplicate the spin in equation 8.4.19

$$\frac{1}{\sqrt{2}} \left(|\uparrow\rangle + |\downarrow\rangle \right) \rightarrow \frac{1}{\sqrt{2}} \left(|\uparrow\rangle + |\downarrow\rangle \right) \frac{1}{\sqrt{2}} \left(|\uparrow\rangle + |\downarrow\rangle \right)$$

$$= \tfrac{1}{2} |\uparrow\rangle \, |\uparrow\rangle + \tfrac{1}{2} |\uparrow\rangle \, |\downarrow\rangle + \tfrac{1}{2} |\downarrow\rangle \, |\uparrow\rangle + \tfrac{1}{2} |\downarrow\rangle \, |\downarrow\rangle$$

$$\tag{8.4.21}$$

The state in equation 8.4.20 is obviously not the same as that in equation 8.4.21. Thus we see that the principle of linearity forbids the existence of quantum Xerox machines. If we could construct Xeroxed quantum states, we would be able to violate the Heisenberg uncertainty principle by a set of measurements on those states.

Chapter 9

The Puzzle of Information Conservation in Black Hole Environments

In 1976 Hawking raised the question of whether information is lost in the process of formation and evaporation of black holes. By information loss, Hawking did not mean the practical loss of information such as would occur in the bomb-in-the-box experiment in Chapter 8. He meant that in a fine grained sense, information would be lost. In other words, the first of the laws of nature described in the Introduction would be violated. The argument was simple and persuasive. It was based on the only available tool of that time, namely local quantum field theory in the fixed background of a black hole. Although Hawking's conclusion is undoubtedly wrong, it played a central role in replacing the old ideas of locality with a new paradigm.

As we have already seen, quantum field theory has a serious defect when it comes to describing systems with horizons. It gives rise to an infinite entropy density on the horizon, instead of the correct Bekenstein–Hawking value of $\frac{c^3}{4G\hbar}$. As we will see, quantum field theory must be replaced with an entirely new paradigm in which the concept of locality is radically altered.

To state the problem, let's begin with a black hole that has been formed during a collapse. The Penrose diagram is shown in Figure 9.1. Following Hawking, we think of the geometry as a background on which we can formulate quantum field theory. Let us concentrate on a theory of massless particles.

According to the principles of quantum mechanics, the evolution of an initial state $|\psi_{in}\rangle$ is governed by a unitary S-matrix, so that the final state is given by

$$|\psi_{out}\rangle \ = \ S \, |\psi_{in}\rangle \tag{9.0.1}$$

One way of stating the principle of information conservation is through the unitarity of S. The point is that a unitary matrix has an inverse, so that

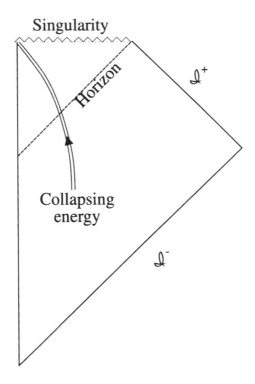

Fig. 9.1   *Penrose diagram of black hole formed by collapse*

in principle the initial state can be recovered from the final state

$$|\psi_{in}\rangle \; = \; S^\dagger \, |\psi_{out}\rangle \qquad\qquad (9.0.2)$$

Now let us consider the S-matrix in a black hole background. The Hilbert space of initial states H_{in} is clearly associated with quanta coming in from $\mathcal{I}^-$. These incoming particles will interact and scatter by means of Feynman diagrams in the black hole background. It is evident from Figure 9.2 that some of the final particles will escape to $\mathcal{I}^+$ and some will be lost behind the horizon. Evidently the final Hilbert space, H_{out} is a tensor product of the states on I^+ and those at the singularity S. Thus

$$\begin{aligned} \mathrm{H}_{in} &= \mathrm{H}_{I^-} \\ \mathrm{H}_{out} &= \mathrm{H}_{I^+} \otimes \mathrm{H}_S \end{aligned} \qquad\qquad (9.0.3)$$

In arguing that the final Hilbert space is a tensor product, Hawking

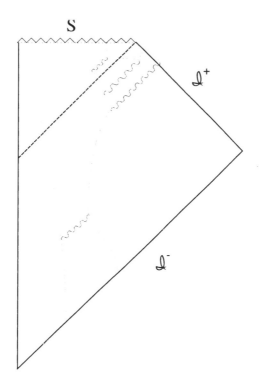

Fig. 9.2   *Feynman diagrams in black hole background*

relied on an important property of quantum field theory: locality. Since every point on the singularity S is space-like relative to every point on $\mathcal{I}^+$, the operators on S all commute with those on $\mathcal{I}^+$. This was central to Hawking's analysis.

Now suppose that the final state is given by an S-matrix which connects H_{in} and H_{out} as in equation 9.0.1. Let us consider the description of the final state from the viewpoint of the observers at $\mathcal{I}^+$. Since they have no access to S, all experiments on the outgoing particles are described by a density matrix

$$\rho_{out} = Tr_{singularity} |\psi_{out}\rangle \langle\psi_{out}| \qquad (9.0.4)$$

where $Tr_{singularity}$ means a trace over the states on the singularity. Thus, in general the observer external to the black hole will see a state characterized by a density matrix.

A simple example might involve a pair of correlated particles which are thrown in from $\mathcal{I}^-$. By correlated, we mean that the two particles wave function is entangled. If one particle ends up on S and the other ends up on $\mathcal{I}^+$, then the final state will be entangled. In this case the state on $\mathcal{I}^+$ will not be pure.

Hawking went on to make arguments that the purity of the state would not be restored if the black hole evaporated. In fact, the only possibilities would seem to be that either information is lost during the entire process of formation and evaporation, *or* the information is restored to the outside world at the very end of the evaporation process, when the singularity is "exposed" at Planckian temperature.

However, we have seen in Chapter 8 that the maximum amount of information that can be hidden in a system is its entropy. Once the entropy of the black hole has evaporated past $\frac{1}{2}$ its original value, it must begin to come out in the emitted radiation. This is a fundamental law of quantum mechanics. By the time the black hole has small mass and entropy, the entanglement entropy of the radiation cannot be larger than the black hole's remaining entropy. Thus, even if all information were emitted at the very end of the evaporation process, a law of nature would be violated from the viewpoint of the external observer. The situation is even worse if the information is not emitted at all.

A final possibility that was advocated by some authors is that black holes never completely evaporate. Instead they end their lives as stable Planck-mass remnants that contain all the lost information. Obviously such remnants would have to have an enormous, or even infinite entropy. Such remnants would be extremely pathological, and we will not pursue that line further.

9.1   A Brick Wall?

There are two more possibilities worth pointing out. One is that the horizon is not penetrable. In other words, from the viewpoints of an in-falling system, the horizon bounces everything out. A freely falling observer would encounter a "brick wall" just above the horizon.

The reason that this was never seriously entertained, especially by relativists, is that it badly violates the equivalence principle. Since the near horizon region of a Schwarzschild black hole is essentially flat space-time, any violent disturbance to a freely falling system would violate the second law of nature in the Introduction. Even more convincing is the fact that the horizon of a black hole formed by a light-like shell forms before the shell gets to the center (see Figures 1.10 and 1.12).

Finally, the quantum Xerox principle closes out the last possibility. The information conservation principle requires all information to be returned to the outside in Hawking radiation. The equivalence principle, on the other hand, requires information to freely pass through the horizon. The quantum Xerox principle precludes both happening. In other words, the horizon cannot duplicate information, and send one copy into the black hole while sending a second copy out. Evidently we have come to an impasse. It seems that some law of nature must break down, at least for some observer.

9.2   Black Hole Complementarity

In its simplest form, black hole complementarity just says that no observer ever witnesses a violation of a law of nature. Thus, for an external observer it says:

A black hole is a complex system whose entropy is a measure of its capacity to store information. It tells us that the entropy is the log of the number of microstates of the degrees of freedom that make up the black hole. It does not tell us what those micro-degrees of freedom are, but it allows us to estimate their number. That number is of order the area of the horizon in Planck units.

Furthermore, it tells us that the micro-degrees of freedom can absorb, thermalize, and eventually re-emit all information in the form of Hawking radiation. At any given time, the fine grained entropy of the radiation field (due to entanglement) cannot exceed the entropy of the black hole. At the end of evaporation, all information is carried off in Hawking radiation.

For a freely falling observer, black hole complementarity tells us that the equivalence principle is respected. This means that as long as the black hole is much larger than the infalling system, the horizon is seen as flat featureless space-time. No high temperatures or other anomolies are encountered.

No obvious contradiction is posed for the external observer, since the infalling observer cannot send reports from behind the horizon. But a potential contradiction can occur for the infalling observer.

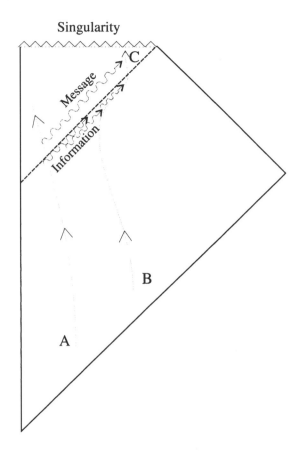

Fig. 9.3   *Information exchange from External to infalling observer*

Consider a black hole, shown in Figure 9.3 along with an infalling system A. System A is assumed to contain some information. According to observations done by A, it passes through the horizon without incident.

Next, consider an observer B who hovers above the black hole monitoring the Hawking radiation. According to assumption, the photons recorded by observer B encode the information carried in by system A. After collecting some information about A (from the Hawking radiation), observer B then jumps into the black hole. We don't actually need observer B to decode the information. All we really need is a mirror outside the black hole horizon to reflect the Hawking radiation back into the black hole.

We now have two copies of the original information carried by A. We can imagine A sending a signal to observer B so that observer B discovers the duplicate information at point C in Figure 9.3. Now we have a contradiction, since observer B has discovered a quantum Xeroxing of information from observer A. If this experiment is possible, then black hole complementarity is not self-consistent. To see why the experiment fails, we need to remember that no information will be emitted until about $\frac{1}{2}$ the entropy of the black hole has evaporated. From equation 5.0.16, this takes a time of order

$$t^* \approx M^3 G^2 \tag{9.2.5}$$

Let us also assume that the observer B hovers above the horizon at a distance at least of the order of the Planck length ℓ_P. In other words, observer B hovers above the stretched horizon. This means that observer B's jump off point must correspond to Rindler coordinates satisfying

$$\omega^* \geq \frac{t^*}{4MG} \approx M^2 G$$

$$\rho^* \geq \ell_P \tag{9.2.6}$$

In terms of the light cone coordinates $x^{\pm} = \rho e^{\pm \omega}$ we have

$$x_*^+ x_*^- > \ell_P^2$$

$$x_*^+ \gtrsim \ell_P \, exp\,(\omega^*) \tag{9.2.7}$$

From Figure 9.4 we can see why there might be a problem with the experiment. Observer B might be forced to the singularity before a message can arrive. In fact, the singularity is given by

$$x^+ x^- = (M\,G)^2 \tag{9.2.8}$$

Thus, observer B will hit the singularity at a point with

$$x^- \lesssim (M\,G)\, e^{-\omega^*} \tag{9.2.9}$$

The implication is that if A is to send a signal that B can receive, it must all occur at $x^- < (M\,G)\, e^{-\omega^*}$. This means that A has a time of order $\Delta t \approx (M\,G)\, e^{-\omega^*}$ to send the message.

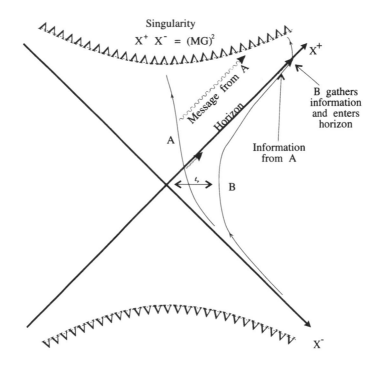

Fig. 9.4   *Resolution of Xerox paradox for observers within horizon*

Now, in the classical theory, there are no limits on how much information can be sent in an arbitrarily small time with arbitrarily small energy. However in quantum mechanics, to send a single bit requires at least one quantum. Since that quantum must be emitted between $x^- = 0$ and $x^- = M G e^{-\omega^*}$, its energy must satisfy

$$E > \frac{1}{MG} e^{\omega^*} \tag{9.2.10}$$

From equation 9.2.6 we see that this energy is exponential in the square of the black hole mass (in Planck units) $E > \frac{e^{M^2 G}}{MG}$. In other words, for observer A to be able to signal observer B before observer B hits the singularity, the energy carried by observer A must be many orders of magnitude larger than the black hole mass. It is obvious that A cannot fit into the horizon, and that the experiment cannot be done.

This example is one of many which show how the constraints of quantum mechanics combine with those of relativity to forbid violations of the complementarity principle.

9.3   Baryon Number Violation

The conservation of baryon number is the basis for the incredible stability of ordinary matter. Nevertheless, there are powerful reasons to believe that baryon number, unlike electric charge, can at best be an approximate conservation law. The obvious difference between baryon number and electric charge is that baryon number is not the source of a long range gauge field. Thus it can disappear without some flux having to suddenly change at infinity.

Consider a typical black hole of stellar mass. It is formed by the collapse of roughly 10^{57} nucleons. Its Schwarzschild radius is about 1 kilometer, and its temperature is 10^{-8} electron volts. It is far too cold to emit anything but very low energy photons and gravitons. As it radiates, its temperature increases, and at some point it is hot enough to emit massive neutrinos and anti-neutrinos, then electrons, muons, and pions. None of these carry off baryon number. It can only begin to radiate baryons when its temperature has increased to about 1 GeV. Using the connection between mass and temperature in equation 4.0.22, the mass of the black hole at this point is about 10^{10} kilograms. This is a tiny fraction of the original black hole mass, and even if it were to decay into nothing but protons, it could produce only about 10^{37} of them. Baryon number must be violated by quantum gravitational effects.

In fact, most modern theories predict baryon violation by ordinary quantum field theoretic processes. As a simplified example, let's suppose there is a heavy scalar particle X which can mediate a transition between an elementary proton and a prositron, as well as between two positrons, as in Figure 9.5. Since the X-boson is described by a real field, it cannot carry any quantum numbers, and the transition evidently violates baryon conservation. The proton could then decay into a positron and an electron-positron pair. Let's also assume that the coupling has the usual Yukawa form

$$g \left[\bar{\psi}_p \, \psi_{e^+} \, X \; + \; \bar{\psi}_{e^+} \, \psi_p \, X \right]$$

where g is a dimensionless coupling.

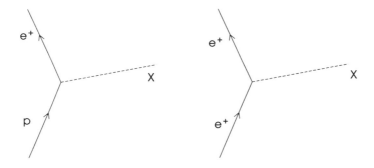

Fig. 9.5   *Feynman diagrams of* $p \rightarrow X e^+$ *and* $e^+ \rightarrow X e^+$

If the mass M_X is sufficiently large, baryon conservation will be a very good symmetry at ordinary energies, just as it is in the real world. The proton will have a lifetime in excess of 10^{32} years.

Now let us consider what happens when a proton falls into a black hole. The baryon number is lost, and will not be radiated back out in the Hawking radiation. The question is: where does the baryon violation take place? One possible answer is that it occurs when the freely falling proton encounters very large curvature invariants as the singularity is approached. From the proton's viewpoint, there is nothing that would stimulate it to decay before that.

On the other hand, from the vantage point of the external observer, the proton encounters enormously high temperatures as it approaches the horizon. Temperatures higher than M_X can certainly excite the proton to decay. So the external observer will conclude that baryon violation can take place at the horizon. Who is right?

The answer that black hole complementarity implies is that they are both right. On the face of it, that sounds absurd. Surely the event of proton decay takes place in some definite place. For a very large black hole, the time along the proton trajectory between horizon crossing and the singularity can be enormous. Thus, it is difficult to understand how there can be an ambiguity.

The real proton propagating through space-time is not the simple structureless bare proton. The interactions cause it to make virtual transitions from the bare proton to a state with an X-boson and a positron. The complicated history of the proton is described by Feynman diagrams such as shown in Figure 9.6. The diagrams make evident that the real proton is

Fig. 9.6   *Proton virtual fluctuation into X e^+ pair*

a superposition of states with different baryon number; in the particular processes shown in Figure 9.6, the intermediate state has vanishing baryon number.

Nothing about virtual baryon non-conservation is especially surprising. As long as the X-boson is sufficiently massive, the rate for real proton decay will be very small, and the proton will be stable for long times.

What is surprising is that the probability to find the proton in a configuration with vanishing baryon number is not small. This probability is closely related to the wave function renormalization constant of the proton, and is of the order

$$Probability \sim \frac{g^2}{4\pi} log \frac{\kappa}{M_X}$$

where κ is the cutoff in the field theory. For example, for $g \sim 1$, κ of the order of the Planck mass, and M_X of the order 10^{16} GeV, the probability that the proton has the "wrong" baryon number is of order unity. This might seem paradoxical, since the proton is so stable.

The resolution of the paradox is that the proton is continuously making extremely rapid transitions between baryon number states. The transitions take place on a time scale of order $\delta t \sim \frac{1}{M_X}$. Ordinary observations of the proton do not see these very rapid fluctuations. The quantity that we normally call baryon number is really the **time averaged** baryon number normalized to unity for the proton.

Now let us consider a proton passing through a horizon, as shown in Figure 9.7. Since the probability that the proton is actually an X, e^+ system is not small, it is not unlikely that when it passes the horizon, its instantaneous baryon number is zero. But it is clear from Figure 9.7 that from the viewpoint of an external observer, this is not a short-lived intermediate state. A fluctuation that is much too rapid to be seen by a low energy observer falling with the proton appears to be a real proton decay lasting to eternity from outside the horizon.

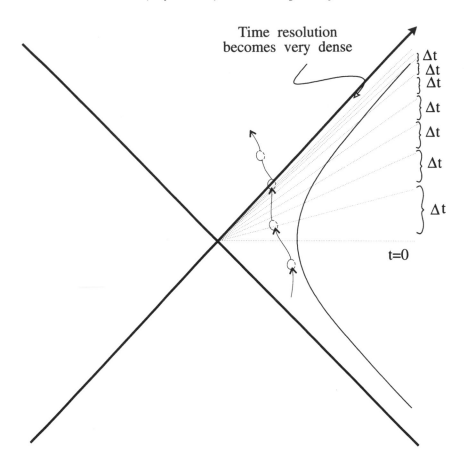

Fig. 9.7   *Proton fluctuations while falling through horizon*

This, of course, is nothing but the usual time dilation near the horizon. As a proton or any other system approaches the horizon, internal oscillations or fluctuations appear to indefinitely slow down, so that a short-lived virtual fluctuation becomes stretched out into a real process.

The process of baryon violation near the horizon should not be totally surprising to the external observer. To him, the proton is falling into a region of increasing temperature. At a proper distance M_X^{-1} from the horizon, the temperature becomes of order M_X. Baryon violating processes are **expected** at that temperature.

An interesting question is whether an observer falling with the proton can observe the baryon number just before crossing the horizon, and send a message to the outside world that the proton has not decayed.

The answer is interesting. In order to make an observation while the proton is in a region of temperature $\gtrsim M_X$ the observer must do so very quickly. In the proton's frame, the time spent in the hot region before crossing the horizon is $\frac{1}{M_X}$. Thus, to obtain a single bit of information about the state of the proton, the observer has to probe it with at least one quantum with energy of order M_X. But such an interaction between the proton and the probe quantum is at high enough energy that it can cause a baryon violating interaction even from the perspective of the proton's frame of reference. Thus the observer cannot measure and report the absence of baryon violation at the horizon without himself causing it.

It is evident from this example that the key to understanding the enormous discrepancies in the complementary description of events lies in understanding the fluctuations of matter at very high frequencies; frequencies so high that the ordinary low energy observer has no chance of detecting them.

Chapter 10

Horizons and the UV/IR Connection

The overriding theme of 20th century physics was the inverse relation between size and momentum/energy. According to conventional relativistic quantum mechanics, a size Δx can be probed with a quantum of energy

$$E_{\Delta x} \approx \frac{\hbar c}{\Delta x} \qquad (10.0.1)$$

But we already know that this trend is destined to be reversed in the physics of the 21st century. This can be seen in many ways. Let's begin with a traditional attempt to study interactions at length scales smaller than the Planck scale. According to conventional thinking, what we need to do is to collide, head on, particles with center of mass energy $E_{\Delta x}$. We expect to discover high energy collision products flying out at all angles. By analyzing the highest energy fragments, we hope to reconstruct very short distance events.

The problem with this thinking is that at energies far above the Planck mass, the collision will create a black hole of mass $\sim E_{\Delta x}$. The interesting short distance effects that we want to probe will be hidden behind a horizon of radius

$$R_S \approx \frac{2G}{c^2} E_{\Delta x}$$

and are inaccessible. To make matters worse, the products of collision will not be high energy particles, but rather low energy Hawking radiation. The typical Hawking particle has energy $\sim \frac{\hbar c}{R_S}$ which decreases with the incident energy. Thus we see that a giant "Super Plancketron" Collider (SPC) would fail to discover fundamental length scales smaller than the Planck scale ℓ_P, no matter how high the energy. In fact, as the energy increases we would be probing ever larger scales

$$\Delta x \sim \frac{2G}{c^2} \frac{E}{c^2}. \tag{10.0.2}$$

We might express this in another form as a kind of space-time uncertainty principle:

$$\Delta x \, \Delta t \sim \frac{2G\hbar}{c^4} \approx \ell_P^2 \tag{10.0.3}$$

This is the simplest example of the ultraviolet/infrared connection that will control the relation between frequency and spatial size. Very high frequency is related to large size scale.

The UV/IR connection is deeply connected to black hole complementarity. As we saw in the previous lecture, the enormous differences in the complementary descriptions of matter falling into a black hole are due to the very different time resolutions available to the complementary observers.

Let's consider further: what happens to a proton falling into the black hole? The proton carries some information with it; its charge, particle type, momentum, location on the horizon through which it falls, etc. From the viewpoint of the external observer, the proton is falling into an increasingly hot region. The proton is like a tiny piece of ice thrown into a tub of very hot water. The only reasonable expectation is that the constituents of the proton "melt" and diffuse throughout the horizon. In fact, in Chapter 7, we saw just such a phenomenon involving a charge falling onto the stretched horizon. More generally, all of the information stored in the incident proton should quickly be spread over the horizon. On the other hand, the observer who follows the proton does not see it spread at all as it falls.

The need to reconcile the complementary descriptions gives us an important clue to the behavior of matter at high frequencies. Following the proton as it falls, the amount of proper time that it has before crossing the horizon tends to zero as the Schwarzschild time tends to infinity. Call the proper time $\Delta\tau$. Then $\Delta\tau$ varies with Schwarzschild time like

$$\Delta\tau \sim \sqrt{8MG\,\delta R} \; e^{-\frac{\Delta t}{4MG}}. \tag{10.0.4}$$

Thus in order to observer the proton before it crosses the horizon, we have to do it in a time which is exponentially small as $t \to \infty$. Thus, what we need in order to be consistent with the thermal spreading of information is now clear. As the proton is observed over shorter and shorter time intervals, the region over which it is localized must grow.

In fact, if we assume that it grows consistently with the uncertainty relation in equation 10.0.3 (substituting $\Delta \tau$ for Δt) we obtain

$$\Delta x \sim \frac{\ell_P^2}{\Delta \tau} \sim \frac{\ell_P^2}{MG} e^{t/4MG}$$

Thus, if the proton size depends on the time resolution according to equation 10.0.3, it will spread over the horizon exponentially fast.

The idea of information spreading as the resolving time goes to zero is very unfamiliar, but it is at the heart of complementarity. It is implicit in the modern idea of the UV/IR connection. Fortunately it is also built into the mathematical framework of string theory.

PART 2

Entropy Bounds and Holography

Chapter 11

Entropy Bounds

11.1 Maximum Entropy

Quantum field theory has too many degrees of freedom to consistently describe a gravitational theory. The main indication that we have seen of this overabundance of degrees of freedom is the fact that the horizon entropy-density is infinite in quantum field theory. The divergence arises from modes very close to the horizon. One might think that this is just an indication that a more or less conventional ultraviolet regulator is needed to render the theory consistent. But the divergent horizon entropy is not an ordinary ultraviolet phenomenon. The modes which account for the divergence are very close to the horizon and would appear to be ultra short distance modes. But they also carry very small Rindler energy and therefore correspond to very long times to the external observer. This is an example of the weirdness of the Ultraviolet/Infrared connection.

A quantitative measure of the overabundance of degrees of freedom in QFT is provided by the *Holographic Principle*. This principle says that there are vastly fewer degrees of freedom in quantum gravity than in any QFT even if the QFT is regulated as, for example, it would be in lattice field theories.

The Holographic Principle is about the counting of quantum states of a system. We begin by considering a large region of space Γ. For simplicity we take the region to be a sphere. Now consider the space of states that describe arbitrary systems that can fit into Γ such that the region outside Γ is empty space. Our goal is to determine the dimensionality of that state-space. Let us consider some preliminary examples.

Suppose we are dealing with a lattice of discrete spins. Let the lattice spacing be a and the volume of Γ be V. The number of spins is then V/a^3

and the number of orthogonal states supported in Γ is given by

$$N_{states} = 2^{V/a^3} \qquad (11.1.1)$$

A second example is a continuum quantum field theory. In this case the number of quantum states will diverge for obvious reasons. We can limit the states, for example by requiring the energy density to be no larger than some bound ρ_{max}. In this case the states can be counted using some concepts from thermodynamics. One begins by computing the thermodynamic entropy density s as a function of the energy density ρ. The total entropy is

$$S = s(\rho)V \qquad (11.1.2)$$

The total number of states is of order

$$N_{states} \sim exp\, S = exp\, s(\rho_{max})V \qquad (11.1.3)$$

In each case the number of distinct states is exponential in the volume V. This is a very general property of conventional local systems and represents the fact that the number of independent degrees of freedom is additive in the volume.

In counting the states of a system the entropy plays a central role. In general entropy is not really a property of a given system but also involves one's state of knowledge of the system. To define entropy we begin with some restrictions that express what we know, for example, the energy within certain limits, the angular momentum and whatever else we may know. The entropy is essentially the logarithm of the number of quantum states that satisfy the given restrictions.

There is another concept that we will call the *maximum entropy*. This *is* a property of the system. It is the logarithm of the total number of states. In other words it is the entropy given that we know nothing about the state of the system. For the spin system the maximum entropy is

$$S_{max} = \frac{V}{a^3} log\, 2 \qquad (11.1.4)$$

This is typical of the maximum entropy. Whenever it exists it is proportional to the **volume**. More precisely it is proportional to the number of simple degrees of freedom that it takes to describe the system.

Let us now consider a system that includes gravity. For definiteness we will take spacetime to be four-dimensional. Again we focus on a spherical region of space Γ with a boundary $\partial\Gamma$. The area of the boundary is A. Suppose we have a thermodynamic system with entropy S that is completely contained within Γ. The total mass of this system can not exceed the mass of a black hole of area A or else it will be bigger than the region.

Now imagine collapsing a spherically symmetric light-like shell of matter with just the right amount of energy so that together with the original mass it forms a black hole which just fills the region. In other words the area of the horizon of the black hole is A. This is shown in Figure 11.1. The

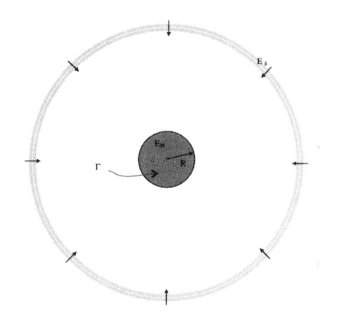

Fig. 11.1 *In-moving (zero entropy) spherical shell of photons*

result of this process is a system of known entropy, $S = A/4G$. But now we can use the second law of thermodynamics to tell us that the original entropy inside Γ had to be less than or equal to $A/4G$. In other words the maximum entropy of a region of space is proportional to its *area* measured in Planck units. Thus we see a radical difference between the number of states of any (regulated) quantum field theory and a theory that includes gravity.

Space-time depiction of horizon formation

Consider the collapsing spherically symmetric shell of light-like energy depicted in Figure 11.1. As the photonic shell approaches the center, the horizon forms prior to the actual crossing of the shell. Inside of the shell, the geometry is Schwarzschild with low curvature (no black hole) prior to the shell crossing the Schwarzschild radius. However, just outside of the shell the geometry becomes increasingly curved as the Schwarzschild radius is approached (see Figure 11.2). The horizon grows until the collapsing

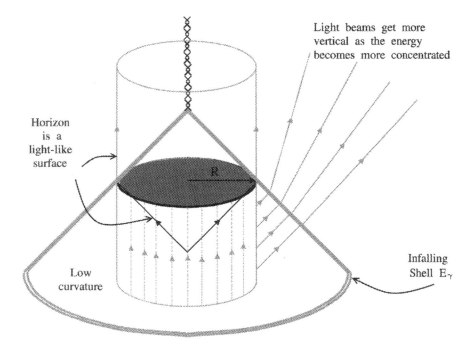

Fig. 11.2 *Space-time depiction of radially in-moving shell of photons*

shell crosses, and a singularity forms at a later time. The energy of the infalling photonic shell E_γ has been tuned such that the collapsing shell crosses the horizon exactly at the radius R in Figure 11.1. However, at that time the system winds up with entropy

$$S = \frac{A}{4G\hbar}.$$

Therefore, unless the second law of thermodynamics is untrue, the entropy of **any** system is limited by

$$S_{max} \leq \frac{A}{4G\hbar}.$$
(11.1.5)

The coarse grained volume in phase space cannot decrease, so this "holographic" limit must be satisfied.

Aside: Scale of Entropy Limit

Example: To get some idea of how big a typical system must be in order to saturate the maximum entropy, consider thermal radiation at a temperature of 1000°K, which corresponds to photons of wavelength $\sim 10^{-5}$ cm. The number of photons N_γ in a volume of radius R satisfies $N_\gamma \sim \frac{V}{\lambda^3} \sim \left(R(cm) \otimes 10^5\right)^3$. Entropy is proportional to the number of photons, and thus one expects

$$S \propto \left(R(cm) \otimes 10^5\right)^3.$$

Compare this with the maximum entropy calculated using the holographic limit

$$S_{max} \approx \left(\frac{R}{l_{Planck}}\right)^2 \cong \left(R(cm) \otimes 10^{33}\right)^2.$$
(11.1.6)

Evidently the maximum entropy will only be saturated for the photon gas when the radius is **huge**, $R \sim 10^{51}$ cm, considerable larger than the observable universe 10^{28} cm.

11.2   Entropy on Light-like Surfaces

So far we have considered the entropy that passes through space-like surfaces. We will see that it is most natural to define holographic entropy bounds on light-like surfaces[3] as opposed to space-like surfaces. Under certain circumstances the entropy bounds of light-like surfaces can be translated to space-like surfaces, but not always. The case described above is one of those cases where a space-like bound is derivable.

Let us start with an example in asymptotically flat space-time. We assume that flat Minkowski coordinates X^+, X^-, x^i can be defined at asymptotic distances. In this chapter we will revert to the usual convention in which X^+ is used as a light cone time variable. We will now define a "light-sheet". Consider the set of all light rays which lie in the surface $X^+ = X_0^+$ in the limit $X^- \to +\infty$. In ordinary flat space this congruence of rays defines a flat three-dimensional light-like surface. In general, they define a light-like surface called a *light sheet*. The light sheet will typically have singular caustic lines, but can be defined in a unique way[4]. When we vary X_0^+ the light sheets fill all space-time except for those points that lie behind black hole horizons.

Now consider a space-time point p. We will assign it light cone coordinates as follows. If it lies on the light sheet X_0^+ we assign it the value $X^+ = X_0^+$. Also if it lies on the light ray which asymptotically has transverse coordinate x_0^i we assign it $x^i = x_0^i$. The value of X^- that we assign will not matter. The two-dimensional x^i plane is called the Screen. Next assume a black hole passes through the light sheet X_0^+. The stretched

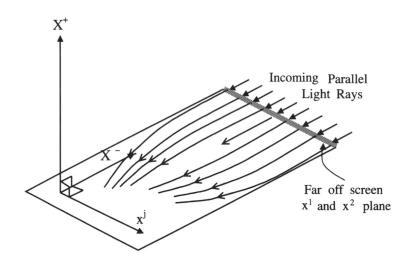

Fig. 11.3 *Light propagating on light-like surface $X^+ = constant$*

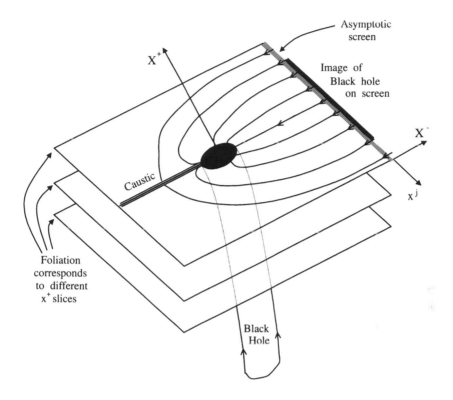

Fig. 11.4   *Family of light rays on fixed* X^+ *surface in presence of black hole*

horizon of the black hole describes a two-dimensional surface in the three-dimensional light sheet as shown in Figure 11.4. Each point on the stretched horizon has unique coordinates X^+, x^i, as seen in Figure 11.5. More generally if there are several black holes passing through the light sheet we can map each of their stretched horizons to the screen in a single valued manner.

Since the entropy of the black hole is equal to $1/4G$ times the area of the horizon we can define an entropy density of $1/4G$ on the stretched horizon. The mapping to the screen then defines an entropy density in the x^i plane, $\sigma(x)$. It is a remarkable fact that $\sigma(x)$ is always less than or equal to $1/4G$.

To prove that $\sigma(x) \leq \frac{1}{4G}$ we make use of the focusing theorem of general relativity. The focusing theorem depends on the positivity of energy and is based on the tendency for light to bend around regions of nonzero energy. Consider a bundle of light rays with cross sectional area α. The light rays

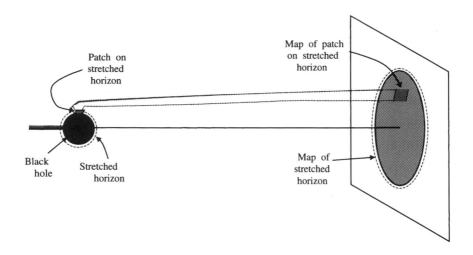

Fig. 11.5   *Image of "stretched horizon" on asymptotic screen*

are parameterized by an affine parameter λ. The focusing theorem says that

$$\frac{d^2\alpha}{d\lambda^2} \leq 0 \qquad\qquad (11.2.7)$$

Consider a bundle of light rays in the light sheet which begin on the stretched horizon and go off to $X^- = \infty$. Since the light rays defining the light sheet are parallel in the asymptotic region $d\alpha/d\lambda \to 0$. The focusing theorem tells us that as we work back toward the horizon, the area of the bundle decreases. It follows that the image of a patch of horizon on the screen is larger than the patch itself. The holographic bound immediately follows.

$$\sigma(x) \leq \frac{1}{4G} \qquad\qquad (11.2.8)$$

This is a surprising conclusion. No matter how we distribute the black holes in three-dimensional space, the image of the entropy on the screen always satisfies the entropy bound equation 11.2.8. An example which helps clarify how this happens involves two black holes. Suppose we try to hide one of them behind the other along the X^- axis, thus doubling the entropy density in the x plane. The bending and focusing of light always acts as in Figures 11.6 to prevent $\sigma(x)$ from exceeding the bound. These considerations lead us to the more general conjecture that for any system,

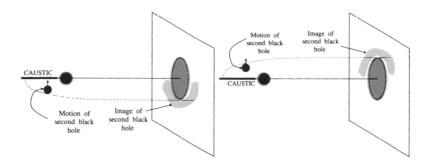

Fig. 11.6   *Initial and later motions and images of second black hole*

when it is mapped to the screen the entropy density obeys the bound in equation 11.2.8.

Thus far we have assumed asymptotically flat boundary conditions. This allowed us to choose the screen so that the light rays forming the light sheet intersect the screen at right angles. Equivalently $da/d\lambda$ equals zero at the screen. We note for future use that the conclusions concerning the entropy bound would be unchanged if we allowed screens for which the light rays were diverging as we move outward, i.e. $da/d\lambda > 0$. However, if we attempt to use screens for which the light rays are converging then the argument fails. This will play an important role in generalizing the holographic bound to more general geometries.

Aside: Apparent motions

Consider a single point particle external to the black hole undergoing motions near a caustic. Examine the projection of those motion upon the screen, demonstrated in Figures 11.7. One sees that due to gravitational lensing, the image of the particle can move at arbitrarily large speeds!

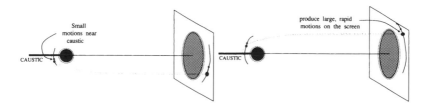

Fig. 11.7   *Initial and later path and image during "slow" motion near caustic*

11.3   Friedman–Robertson–Walker Geometry

The holographic bound can be generalized to *flat* F.R.W. geometries, where it is called the Fischler–Susskind (FS) bound[5] and to more general geometries by Bousso[6]. First we will review the F.R.W. case. Consider the general case of $d + 1$ dimensions. The metric has the form

$$d\tau^2 = dt^2 - a(t)^2 dx^m dx^m \qquad (11.3.9)$$

where the index m runs over the d spatial directions. The function $a(t)$ is assumed to grow as a power of t.

$$a(t) = a_0 t^p \qquad (11.3.10)$$

Let's also make the usual simplifying cosmological assumptions of homogeneity. In particular we assume that the spatial entropy density (per unit d volume) is homogeneous. Later, we will relax these assumptions.

At time t we consider a spherical region Γ of volume V and area A. The boundary $(d-1)$-sphere, $\partial\Gamma$, will play the same role as the screen in the previous discussion. The light sheet is now defined by the backward light cone formed by light rays that propagate from $\partial\Gamma$ into the past. (See Figure 11.8.)

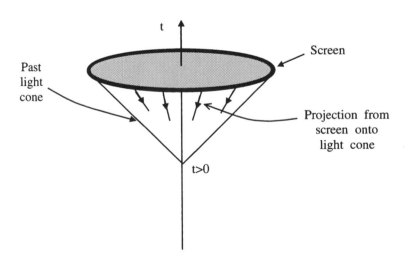

Fig. 11.8 *Holographic surface for calculating entropy bound with a spherical surface as the screen*

As in the previous case the holographic bound applies to the entropy passing through the light sheet. This bound states that the total entropy passing through the light sheet does not exceed $A/4G$. The key to a proof is again the focusing theorem. We observe that at the screen the area of the outgoing bundle of light rays is increasing as we go to later times. In other words the light sheet has positive expansion into the future and negative expansion into the past. The focusing theorem again tells us that if we map the entropy of black holes passing through the light sheet to the screen, the resulting density satisfies the holographic bound. It is believed that the bound is very general.

It is now easy to see why we concentrate on light sheets instead of space-like surfaces. Obviously if the spatial entropy density is uniform and we choose Γ big enough, the entropy will exceed the area. However if Γ is larger than the particle horizon at time t the light sheet is not a cone, but rather a truncated cone as in Figure 11.9, which is cut off by the big bang at $t = 0$. Thus a portion of the entropy present at time t never passed through the light sheet. If we only count that portion of the entropy which did pass through the light sheet, it will scale like the area A. We will return to the question of space-like bounds after discussing Bousso's generalization[6] of the FS bound.

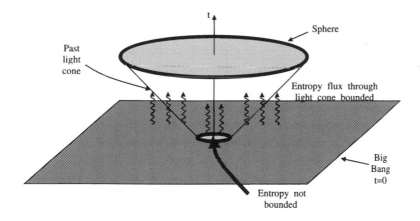

Fig. 11.9 *Region of space causally connected to particle horizon*

Test: Is the observed horizon entropy bounded by its area?

We will check whether the observed particle horizon satisfies the entropy bound:

$$S_{Horizon} \leq \frac{A_{Horizon}}{4\,G} \text{ ???}$$

We can check by recognizing that the entropy is primarily given by the number of black body cosmic background photons, $N_\gamma \approx 10^{90}$. The proper size of the horizon is approximately given by 10^{18} (light) seconds, and the Planck time is approximately 10^{-43} seconds. This gives a proper size for the horizon of about 10^{61} Planck units. We can use these numbers to compare the cosmic entropy with the area of the horizon:

$$S_{Horizon} \leq^{???} \frac{A_{Horizon}}{4\,G}$$

$$10^{90} \leq \left(10^{61}\right)^2 = 10^{122}$$

We find this inequality to definitely be true *today*.

Next, using the F.R.W. geometry, we will determine if the entropy per horizon-area contained within the particle horizon is increasing or decreasing. Let $R_{Horizon}$ represent the coordinate size of the particle horizon (Figure 11.10), and d represent the number of spatial dimensions. Let σ be the entropy volume density, so that

$$S \sim \sigma\, R_{Horizon}^d$$

$$d\tau^2 = dt^2 - a^2(t) \sum_{j=1}^{d} dx^j\, dx^j$$

This means that the proper size of the horizon is given by $a(t)R_{Horizon}$. We want to check whether

$$S \sim \sigma\, R_{Horizon}^d \, <^? \, \frac{\left(a(t)\, R_{Horizon}\right)^{d-1}}{4\,G} \tag{11.3.11}$$

An outgoing light ray (null geodesic) which would generate the particle horizon satisfies $dt = a(t)dx$, which gives the form for the time dependence

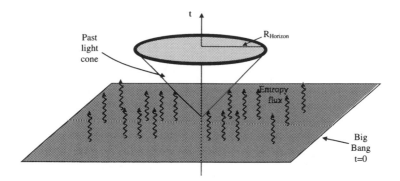

Fig. 11.10   *Particle Horizon*

of the size of the particle horizon:

$$R_{Horizon}(t) = \int\limits_0^t \frac{dt'}{a(t')}$$ (11.3.12)

If we assume the form $a(t) = a_o t^p$ then the particle horizon evolves according to the formula

$$R_{Horizon}(t) = \frac{t^{1-p}}{a_o}.$$ (11.3.13)

Therefore for the entropy bound to continue to be valid, the time dependence must satisfy $t^{(1-p)d} < t^{d-1}$. This bounds the expansion rate coefficient to satisfy

$$p > \frac{1}{d}.$$ (11.3.14)

One sees that if the expansion rate is too slow, then the coordinate volume will grow faster than the area, and the entropy bound will eventually be contradicted.

Suppose that the matter in the F.R.W. cosmology satisfies the equation of state

$$P \;=\; w\,u$$ (11.3.15)

where P is the pressure and u is the energy density and w is a constant. Given w and the scale factor for the expansion p, one can use the Einstein field equation to calculate a relationship between them:

$$p = \frac{2}{d(1+w)} > \frac{1}{d} \qquad (11.3.16)$$

We see that the number of spatial dimensions cancels and that the entropy bound is satisfied so long as

$$w \leq 1. \qquad (11.3.17)$$

This is an interesting result. Recall that the speed of sound within a medium is given by

$$v_s^2 = \tfrac{\partial P}{\partial u} = w$$

Therefore, in the future, the bound will **always** be satisfied, since the speed of sound is always less than the speed of light. The relation satisfied by the scale factor $v_s^2 = w \leq 1$ is just the usual causality requirement. As one moves forward in time, the entropy bound then becomes **more** satisfied, not less.

Next, go back in time using the black body radiation background as the dominant entropy. Using a decoupling time $t_{decoupling} \sim 10^5$ years (when the background radiation fell out of equilibrium with the matter) and extrapolating back using the previously calculated entropy relative to the bound, one gets

$$\frac{S}{A/(4G)} = 10^{-28} \left[\frac{t_{decoupling}}{t} \right]^{\frac{1}{2}} \qquad (11.3.18)$$

The entropy bound $S = \frac{A}{4G}$ is reached when

$$\left[\frac{t_{decoupling}}{t} \right]^{\frac{1}{2}} = 10^{28} \Rightarrow t = \frac{t_{decoupling}}{10^{56}} \sim 10^{-44} \, \text{sec} \qquad (11.3.19)$$

This time is comparable to the Planck time (by coincidence??). Therefore the entropy bound is not exceeded after the Planck time.

11.4   Bousso's Generalization

Consider an arbitrary cosmology. Take a space-like region Γ bounded by the space-like boundary $\partial\Gamma$. At any point on the boundary we can construct four light rays that are perpendicular to the boundary[6]. We will call these the four *branches*. Two branches go toward the future. One of them is composed of outgoing rays and the other is ingoing. Similarly two branches go to the past. On any of these branches a light ray, together with its neighbors define a positive or negative expansion as we move away from the boundary. In ordinary flat space-time, if $\partial\Gamma$ is convex the outgoing (ingoing) rays have positive (negative) expansion. However in non-static universes other combinations are possible. For example in a rapidly contracting universe both future branches have negative expansion while the past branches have positive expansion.

If we consider general boundaries the sign of the expansion of a given branch may vary as we move over the surface. For simplicity we restrict attention to those regions for which a given branch has a unique sign. We can now state **Bousso's rule**: *From the boundary $\partial\Gamma$ construct all light sheets which have negative expansion as we move away. These light sheets may terminate at the tip of a cone or a caustic or even a boundary of the geometry. Bousso's bound states that the entropy passing through these light sheets is less than $A/4G$ where A is the boundary of $\partial\Gamma$.*

To help visualize how Bousso's construction works we will consider spherically symmetric geometries and use Penrose diagrams to describe them. The Penrose diagram represents the radial and time directions. Each point of such a diagram really stands for a 2-sphere (more generally a $(d-1)$-sphere). The four branches at a given point on the Penrose diagram are represented by a pair of 45 degree lines passing through that point. However we are only interested in the branches of negative expansion. For example in Figure 11.11 we illustrate flat space-time and the negative expansion branches of a typical local 2-sphere. In general as we move around in the Penrose diagram the particular branches which have negative expansion may change. For example if the cosmology initially expands and then collapses, the outgoing future branch will go from positive to negative expansion. Bousso introduced a notation to indicate this. The Penrose diagram is divided into a number of regions depending on which branches have negative expansion. In each region the negative expansion branches are indicated by their directions at a typical point. Thus in Figure 11.12 we draw the Penrose–Bousso (PB) diagram for a positive curvature, matter dominated universe that begins with a bang and ends with a crunch. It consists of four distinct regions.

In Region I of Figure 11.12 the expansion of the universe causes both past branches to have negative expansion. Thus we draw light surfaces into the past. These light surfaces terminate on the initial boundary of the

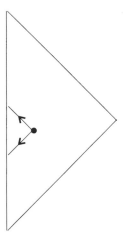

Fig. 11.11   *Negative expansion branches of 2-sphere in flat space-time*

geometry and are similar to the truncated cones that we discussed in the flat F.R.W. case. The holographic bound in this case says that the entropy passing through either backward light surface is bounded by the area of the 2-sphere at point p. Bousso's rule tells us nothing in this case about the entropy on space-like surfaces bounded by p.

Now move on to Region II. The relevant light sheets in this region begin on the 2-sphere q and both terminate at the spatial origin. These are untruncated cones and the entropy on both of them is holographically bounded. There is something new in this case. We find that the entropy is bounded on a future light sheet. Now consider a space-like surface bounded by q and extending to the spatial origin (shown in Figure 11.13). It is evident that any matter which passes through the space-like surface must also pass through the future light sheet. By the second law of thermodynamics the entropy on the space-like surface can not exceed the entropy on the future light sheet. Thus in this case the entropy in a space-like region can be holographically bounded. Therefore, one condition for a space-like bound is that the entropy is bounded by a corresponding future light sheet. With this in mind we return to Region I. For Region I there is no future bound, and therefore the entropy is not bounded on space-like regions with boundary p.

In Region III the entropy bounds are both on future light sheets. Nevertheless there is no space-like bound. The reason is that not all matter

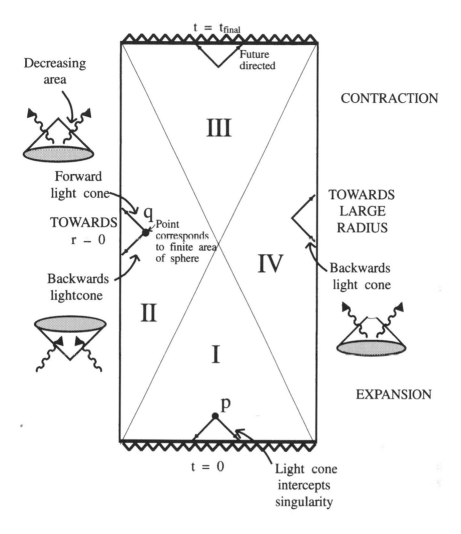

Fig. 11.12 *Penrose–Bousso diagram for matter dominated universe*

which pass through space-like surfaces are forced to pass through the future light sheets.

Region IV is identical to Region II with the spatial origin being replaced by the diametrically opposed antipode. Thus we see that there are light-like bounds in all four regions but only in II and IV are there holographic bounds on space-like regions. (See Figure 11.13.) Figure 11.14 demonstrates a region that does not satisfy an entropy bound for this cosmology.

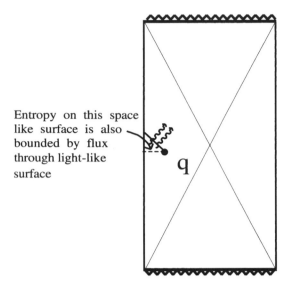

Entropy on this space
like surface is also
bounded by flux
through light-like
surface

q

Fig. 11.13   *Bounding surfaces for inflowing entropy*

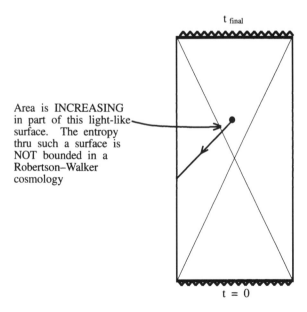

t final

Area is INCREASING
in part of this light-like
surface.   The entropy
thru such a surface is
NOT bounded in a
Robertson–Walker
cosmology

t = 0

Fig. 11.14   *Light-like surface which does not satisfy entropy bound*

11.5   de Sitter Cosmology

de Sitter space holds special interest because of its close connection with observational cosmology. The model of cosmology which is presently gaining the status of a "standard model" involves de Sitter space in two ways. First it is believed that the early universe underwent an epoch of rapid inflation during which the space-time geometry was very close to de Sitter. The inflationary theory is more than two decades old and by now has been well tested.

More recently de Sitter space has entered cosmology as the likely candidate for the final fate of the universe. The history of the universe seems to be a transition from an early de Sitter epoch in which the vacuum energy or cosmological constant was very large to a late de Sitter phase characterized by a very small but not zero cosmological constant.

de Sitter space is the solution of Einstein's field equations with a positive cosmological constant that exhibits maximal symmetry. Four-dimensional de Sitter space may be defined by embedding it in $(4 + 1)$ dimensional flat Minkowski space. It is the hyperboloid given by

$$\sum_{i=1}^{4} (x^i)^2 - (x^0)^2 = R^2. \tag{11.5.20}$$

The radius of curvature R is related to the cosmological constant, λ.

$$R^2 = \frac{3}{G\lambda} \tag{11.5.21}$$

de Sitter space can also be written in the form

$$d\tau^2 = dt^2 - a(t)^2 d\Omega_3^2 \tag{11.5.22}$$

where $d\Omega_3^2$ is the metric of a unit 3-sphere, and the scale factor a is given by

$$a(t) = R\cosh(t/R). \tag{11.5.23}$$

For our purposes we want to put this in a form that will allow us to read off the Penrose diagram. To that end define the conformal time T by

$$dT = dt/a(t). \tag{11.5.24}$$

One easily finds that the geometry has the form

$$d\tau^2 = (a(t))^2 (dT^2 - d\Omega_3^2). \tag{11.5.25}$$

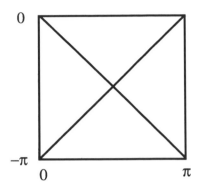

Fig. 11.15   *Penrose diagram for de Sitter space-time*

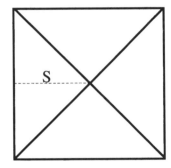

Fig. 11.16   *Extremal space-like surface in de Sitter space-time*

Furthermore when t varies between $\pm\infty$, the conformal time ($T = tan^{-1}\left(tanh(t/R)\right) - \frac{\pi}{2}$) varies from $-\pi$ to 0. Since the polar angle on the sphere also varies from 0 to π the Penrose diagram is a square as in Figure 11.15.

Once again the Bousso construction divides the Penrose diagram into four quadrants. However, the contracting light sheets in the upper and lower quadrants are oriented oppositely to the usual F.R.W. case. The reason is that the geometry rapidly expands as we move toward the upper and lower boundary of the de Sitter space.

Of particular interest is the bound on a space-like surface beginning at the center of the Penrose diagram and extending to the left edge as in Figure 11.16. We leave it as an exercise to show that the entropy is

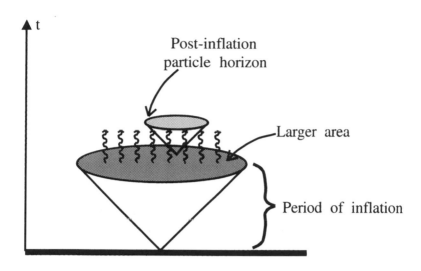

Fig. 11.17   *Inflationary universe*

bounded by the area of the 2-sphere at the center of the diagram. In fact
the maximum entropy on any such surface is given by

$$S = \frac{4\pi R^2}{4G}. \tag{11.5.26}$$

de Sitter space has a special importance because of its role during the
early evolution of the universe. According to the inflationary hypothesis,
the universe began with a large vacuum energy which mimicked the effects
of a positive cosmological constant. During that period the geometry was de
Sitter space. But then at some time the vacuum energy began to decrease
and the universe made a transition to an F.R.W. universe. The transition
was accompanied by the production of a large amount of entropy, and is
called *reheating*.

Inflationary cosmology is illustrated in Figure 11.17. In order to construct the Penrose–Bousso diagram we begin by drawing Penrose diagrams
for both de Sitter space and either radiation or matter dominated F.R.W.
The Penrose–Bousso diagram for de Sitter space is shown in Figure 11.18.
In order to describe inflationary cosmology we must terminate the de Sitter
space at some late time and attach it to a conventional F.R.W. space as in
Figure 11.19. The dotted line where the two geometries are joined is the
reheating surface where the entropy of the universe is created.

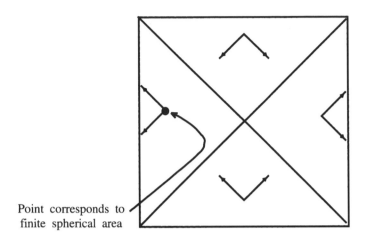

Point corresponds to
finite spherical area

Fig. 11.18　*Buosso wedges for de Sitter geometry*

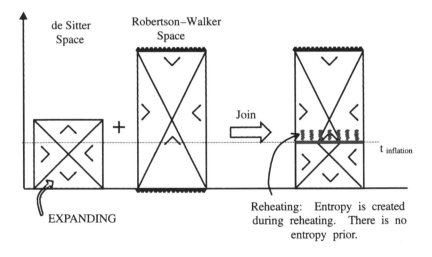

de Sitter Space

Robertson–Walker Space

Join

EXPANDING

$t_{\text{inflation}}$

Reheating: Entropy is created
during reheating. There is no
entropy prior.

Fig. 11.19　*Joining of inflationary and post-inflationary geometries*

Let us focus on the point p in Figure 11.20. It is easy to see that in an ordinary inflationary cosmology p can be chosen so that the entropy on the space-like surface $p - q$ is bigger than the area of p. However Bousso's rule applied to point p only bounds the entropy on the past light sheet.

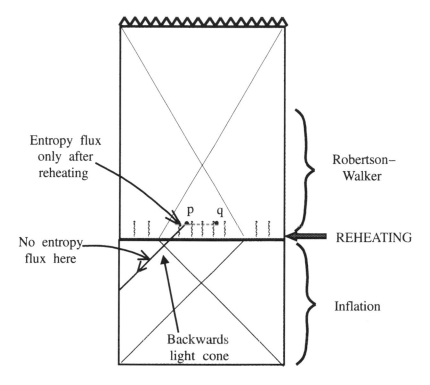

Fig. 11.20 *Entropy bound on spherical region causally connected to inflationary period*

In this case most of the newly formed entropy on the reheating surface is not counted since it never passed through the past light sheet. Typical inflationary cosmologies can be studied to see that the past light sheet bound is not violated.

11.6 Anti de Sitter Space

de Sitter space is important because it may describe the early and late time behavior of the real universe. Anti de Sitter space is important for an entirely different reason. It is the background in which the holographic principle is best understood. In Chapter 12 the geometry of AdS will be reviewed, but for our present purpose all we need are the properties of its PB diagram.

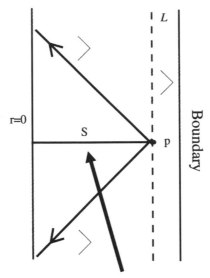

ANY spacelike surface S will have
its entropy bounded by the area

Fig. 11.21   *Static surface of large area in AdS space*

AdS space is the vacuum of theories with a negative cosmological constant. The space-time in appropriate coordinates is static but unlike de Sitter space the static coordinates cover the entire space. The vacuum of AdS is a genuine zero temperature state.

The PB diagram consists of an infinite strip bounded on the left by the spatial origin and on the right by a boundary. The PB diagram consists of a single region in which both negative expansion light sheets point toward the origin. Let us consider a static surface of large area A far from the spatial origin. The surface is denoted by the dotted vertical line L in Figure 11.21. We will think of L as an infrared cutoff. Consider an arbitrary point p on L. Evidently Bousso's rules bound the entropy on past and future light sheets bounded by p. Therefore the entropy on any space-like surface bounded by p and including the origin is also holographically bounded. In other words the entire region to the left of L can be foliated with space-like surfaces such that the maximum entropy on each surface is $A/4G$.

AdS space is an example of a special class of geometries which have time-like Killing vectors and which can be foliated by space-like surfaces that satisfy the Holographic bound. These two properties imply a very far reaching conclusion. All physics taking place in such backgrounds (in the interior of the infrared cutoff L) must be described in terms of a Hamiltonian that acts in a Hilbert space of dimensionality

$$N_{states} = exp(A/4G) \qquad (11.6.27)$$

The holographic description of AdS space is the subject of the next chapter.

Chapter 12

The Holographic Principle and Anti de Sitter Space

12.1 The Holographic Principle

As we have seen in Chapter 11, the number of possible quantum states in a region of flat space is bounded by the exponential of the area of the region in Planck units. That fact together with the Ultraviolet/Infrared connection and Black Hole Complementarity has led physics to an entirely new paradigm about the nature of space, time and locality. One of the elements of this paradigm is the Holographic Principle and its embodiment in AdS space.

Let us consider a region of flat space Γ. We have seen that the maximum entropy of all physical systems that can fit in Γ is proportional to the area of the boundary $\partial\Gamma$, measured in Planck units. Typically, as in the case of a lattice of spins, the maximum entropy is a measure of the number of simple degrees of freedom* that it takes to completely describe the region. This is almost always proportional to the volume of Γ. The exception is gravitational systems. The entropy bound tells us that the maximum number of non-redundant degrees of freedom is proportional to the area. For a large macroscopic region this is an enormous reduction in the required degrees of freedom. In fact if the linear dimensions of the system is of order L then the number of degrees of freedom per unit volume scales like $1/L$ in Planck units. By making L large enough we can make the degrees of freedom arbitrarily sparse in space. Nevertheless we must be able to describe microscopic processes taking place anywhere in the region. One way to think of this is to imagine the degrees of freedom of Γ as living on $\partial\Gamma$ with an area density of no more than ~ 1 degree of freedom per

---

*By a simple degree of freedom we mean something like a spin or the presence or absence of a fermion. A simple degree of freedom represents a single bit of information.

Planck area. The analogy with a hologram is obvious; three-dimensional space described by a two-dimensional hologram at its boundary! That this is possible is called the Holographic Principle.

What we would ideally like to do is to have a solution of Einstein's equations that describes a ball of space with a spherical boundary and then to count the number of degrees of freedom. Even better would be to construct a description of the region in terms of a boundary theory with a limited density of degrees of freedom. Ordinarily, it does not make sense to consider a ball-like region with a boundary in the general theory of relativity. But there is one special situation which is naturally ball-like. It occurs when there is a negative cosmological constant; Anti de Sitter space. Thus AdS is a natural framework in which to study the Holographic Principle.

12.2   AdS Space

We saw in Chapter 11 that AdS space enjoys certain properties which make it a natural candidate for a holographic Hamiltonian description. In this lecture we will describe a very precise version of AdS "Holography" which grew out of the mathematics of string theory. The remarkable precision is due to the unusually high degree of symmetry of the theory which includes a powerful version of supersymmetry. However we will downplay the mathematical aspects of the theory and concentrate on those physical principles which are likely to be general.

The particular space that we will be interested in is not simple 5-dimensional AdS but rather $AdS(5) \otimes S(5)$[7][8][9]. This is a 10-dimensional product space consisting of two factors, the 5-dimensional AdS and a 5-sphere $S(5)$. Why the $S(5)$? The reason involves the high degree of supersymmetry enjoyed by superstring theory. Generally the kind of supergravity theories that emerge from string theory don't have cosmological constants. But by bending some of the directions of space into compact manifolds it becomes possible to generate a cosmological constant for the resulting lower dimensional Kaluza–Klein type theory. From a conceptual point of view the extra internal 5-sphere is not important. From a mathematical point of view it is essential if we want to be able to make precision statements.

We will begin with a brief review of AdS geometry. For our purposes 5-dimensional AdS space may be considered to be a solid 4-dimensional spatial ball times the infinite time axis. The geometry can be described by dimensionless coordinates t, r, Ω where t is time, r is the radial coordinate ($0 \leq r < 1$) and Ω parameterizes the unit 3-sphere. The geometry has uniform curvature R^{-2} where R is the radius of curvature. The metric we will use is

$$d\tau^2 = \frac{R^2}{(1-r^2)^2} \left\{ (1+r^2)^2 dt^2 - 4dr^2 - 4r^2 d\Omega^2 \right\} \tag{12.2.1}$$

There is another form of the metric which is in common use,

$$d\tau^2 = \frac{R^2}{y^2} \left\{ dt^2 - dx^i dx^i - dy^2 \right\} \tag{12.2.2}$$

where i runs from 1 to 3.

The metric in equation 12.2.2 is related to 12.2.1 in two different ways. First of all it is an approximation to equation 12.2.1 in the vicinity of a point on the boundary at $r = 1$. The 3-sphere is replaced by the flat tangent plane parameterized by x^i and the radial coordinate is replaced by y, with $y = (1 - r)$.

The second way that equations 12.2.1 and 12.2.2 are related is that 12.2.2 is the exact metric of an incomplete patch of AdS space. A time-like geodesic can get to $y = \infty$ in a finite proper time so that the space in equation 12.2.2 is not geodesically complete. It has a horizon at $y = \infty$. When interpreted in this manner, time coordinates appearing in 12.2.1 and 12.2.2 are not the same.

The metric 12.2.2 may be expressed in terms of the coordinate $z = 1/y$.

$$d\tau^2 = R^2 \left\{ z^2 (dt^2 - dx^i dx^i) - \frac{1}{z^2} dz^2 \right\} \tag{12.2.3}$$

In this form it is clear that there is a horizon at $z = 0$ since the time–time component of the metric vanishes there. The boundary is at $z = \infty$.

To construct the space $AdS(5) \otimes S(5)$ all we have to do is define 5 more coordinates ω_5 describing the unit 5-sphere and add a term to the metric

$$ds_5^2 = R^2 d\omega_5^2 \tag{12.2.4}$$

Although the boundary of AdS is an infinite proper distance from any point in the interior of the ball, light can travel to the boundary and back in a finite time. For example, it takes a total amount of (dimensionless) time $t = \pi$ for light to make a round trip from the origin at $r = 0$ to the boundary at $r = 1$ and back. For all practical purposes AdS space behaves like a finite cavity with reflecting walls. The size of the cavity is of order R. In what follows we will think of the cavity size R as being much larger than any microscopic scale such as the Planck or string scale.

Supplement on Properties of AdS metric

1) The point $r = 0$ is the center of the Anti de Sitter space and r varies from 0 to 1 in the space. A radial null geodesic satisfies $\left(1 + r^2\right)^2 dt^2 = 4dr^2$, which means that a light beam will traverse the infinite proper distance in r from 0 to 1 back to 0 in a round trip time given by π, which makes the space causally finite.
2) The metric is singular at $r = 1$ in **all** components. A unit coordinate time interval corresponds to increasingly large proper time intervals.
3) Near $r = 1$, the metric is approximately conformal, which means that light rays move at 45° angles near the boundary.

$$ds^2 \cong \tfrac{4R^2}{(1-r^2)^2} \left\{ -dt^2 + dr^2 + r^2 d\Omega_{D-2}^2 \right\}$$

Light rays move slower (by a factor of 2) near the center of the Anti de Sitter space.
4) Generally, the spatial metric is that of a uniformly (negatively) curved space, a *hyperbolic plane* (or the *Poincaré disk*).

12.3   Holography in AdS Space

We will refer to $AdS(5) \otimes S(5)$ as the bulk space and the 4-dimensional boundary of AdS at $y = 0$ as the boundary. According to the Holographic Principle we should be able to describe everything in the bulk by a theory whose degrees of freedom can be identified with the boundary at $y = 0$. However the Holographic Principle requires more than that. It requires that the boundary theory has no more than 1 degree of freedom per Planck area. To see what this entails, let us compute the area of the boundary. From equation 12.2.1 we see that the metric diverges near the boundary.

Later we will regulate this divergence by moving in a little way from $y = 0$, but for the time being we can assume that the number of degrees of freedom per unit *coordinate* area is infinite. That suggests that the boundary theory might be a quantum field theory, and that is in fact the case.

Another important fact involves the symmetry of AdS. Let us consider the metric in the form 12.2.2. It is obvious that the geometry is invariant under ordinary Poincaré transformations of the 4-dimensional Minkowski coordinates t, x^i. In addition there is a "dilatation" symmetry

$$t \to \lambda t$$

$$x^i \to \lambda x^i$$

$$y \to \lambda y \tag{12.3.5}$$

On the other hand if we consider the representation of AdS in 12.2.1 we can see additional symmetry. For example the rotations of the sphere Ω are symmetries. The full symmetry group of $AdS(5)$ is the group $O(4|2)$. In addition there is also the symmetry $O(6)$ associated with rotations of the internal 5-sphere.

Since our goal is a holographic boundary description of the physics in the bulk spacetime it is very relevant to ask how the symmetries act on the boundary of AdS. Obviously, the 4-dimensional Poincaré symmetry acts on the boundary straightforwardly. The dilatation symmetry also acts as a simple dilatation of the coordinates t, x. All of the transformations act on the boundary as conformal transformations which preserve light-like directions on the boundary. In fact the full AdS symmetry group, when acting on the boundary at $y = 0$ is precisely the conformal group of 4-dimensional Minkowski space.

The implication of this symmetry of the boundary is that the holographic boundary theory must be invariant under the conformal group. This together with the fact that the boundary has an infinite (coordinate) density of degrees of freedom suggests that the holographic theory is a Conformal Quantum Field Theory, and so it is.

As we mentioned, $AdS(5) \otimes S(5)$ is a solution of the 10-dimensional supergravity that describes low energy superstring theory. Indeed the space has more symmetry than just the conformal group and the $O(6)$ symmetry of the internal 5-sphere. The additional symmetry is the so-called $\mathcal{N} = 4$ supersymmetry. This symmetry must also be realized by the holographic theory. All of this leads us to the remarkable conclusion that quantum gravity in $AdS(5) \otimes S(5)$ should be exactly described by an appropriate superconformal Lorentz invariant quantum field theory associated with the AdS boundary.

In order to have a benchmark for the counting of degrees of freedom in $AdS(5) \otimes S(5)$ imagine constructing a cutoff field theory in the bulk. A conventional cutoff would involve a microscopic length scale such as the 10-dimensional Planck length l_p. One way to do this would be to introduce a spatial lattice in 9-dimensional space. It is not generally possible to make a regular lattice, but a random lattice with an average spacing l_p is possible. We can then define a simple theory such as a Hamiltonian lattice theory on the space. In order to count degrees of freedom we also need to regulate the area of the boundary of AdS which is infinite. To do so we introduce a surface L at $r = 1 - \delta$. The total 9-dimensional spatial volume in the interior of L is easily computed using the metric 12.2.2, and is seen to be critically divergent.

$$V(\delta) \sim \frac{R^9}{\delta^3}.$$ (12.3.6)

The number of bulk lattice sites and therefore the number of degrees of freedom is

$$\frac{V}{l_p^9} \sim \frac{1}{\delta^3} \frac{R^9}{l_p^9}$$ (12.3.7)

In such a theory we also will find that the maximum entropy is of the same order of magnitude. However the Holographic Principle suggests that this entropy is overestimated.

The holographic bound discussed in Chapter 11 requires the maximum entropy and the number of degrees of freedom to be of order

$$S_{max} \sim \frac{A}{l_p^8}$$ (12.3.8)

where A is the 8-dimensional area of the boundary L. This is also easily computed. We find

$$S_{max} \sim \frac{1}{\delta^3} \frac{R^8}{l_p^8}$$ (12.3.9)

In other words when R/l_p becomes large the holographic description requires a reduction in the number of independent degrees of freedom by a factor l_p/R. To say it slightly differently, the Holographic Principle implies a complete description of all physics in the bulk of a very large AdS space in terms of only l_p/R degrees of freedom per spatial Planck volume. Nonetheless the theory must be able to describe microscopic events in the bulk even when R becomes extremely large.

12.4   The AdS/CFT Correspondence

The search for a holographic description of $AdS(5) \otimes S(5)$ is considerably narrowed by the symmetries. In fact there is only one known class of systems with the appropriate $\mathcal{N} = 4$ supersymmetry; the $SU(N)$ Supersymmetric Yang–Mills (SYM) theories.

The correspondence between gravity or its string theoretic generalization in $AdS(5) \otimes S(5)$ and Super Yang–Mills (SYM) theory on the boundary is the subject of a vast literature. We will only review some of the salient features. The correspondence states that there is a complete equivalence between superstring theory in the bulk of $AdS(5) \otimes S(5)$ and $\mathcal{N} = 4$, $3 + 1$-dimensional, $SU(N)$, SYM theory on the boundary of the AdS space[7][8][9]. In these lectures SYM theory will always refer to this particular version of supersymmetric gauge theory, $\mathcal{N}$ represents the number of supersymmetries, and N is the dimension of the Yang–Mills gauge theory.

It is well known that SYM is conformally invariant and therefore has no dimensional parameters. It will be convenient to define the theory to live on the boundary parametrized by the dimensionless coordinates t, Ω or t, x. The corresponding momenta are also dimensionless. In fact we will use the convention that all SYM quantities are dimensionless. On the other hand the bulk gravity theory quantities such as mass, length and temperature carry their usual dimensions. To convert from SYM to bulk variables, the conversion factor is R. Thus if E_{SYM} and M represent the energy in the SYM and bulk theories

$$E_{SYM} = MR.$$

Similarly bulk time intervals are given by multiplying the t interval by R.

There is one question that may be puzzling to the reader. Since $AdS(5) \otimes S(5)$ is a 10-dimensional spacetime one might think that its boundary is $(8 + 1)$ dimensional. But there is an important sense in which it is $3 + 1$ dimensional. To see this let us Weyl rescale the metric by a factor $\frac{R^2}{(1-r^2)^2}$ so that the rescaled metric at the boundary is finite. The new metric is

$$dS^2 = \{(1+r^2)^2 dt^2 - 4dr^2 - 4r^2 d\Omega^2\} + \{(1-r^2)^2 d\omega_5^2\} \qquad (12.4.10)$$

Notice that the size of the 5-sphere shrinks to zero as the boundary at $r = 1$ is approached. The boundary of the geometry is therefore 3+1 dimensional.

Let us return to the correspondence between the bulk and boundary theories. The 10-dimensional bulk theory has two dimensionless parameters. These are:

1) The radius of curvature of the AdS space measured in string units R/ℓ_s. Alternately we could measure R in 10-dimensional Planck units. The relation between string and Planck lengths is given by

$$g^2 \ell_s^8 = l_p^8$$

2) The dimensionless string coupling constant g.

The string coupling constant and length scale are related to the 10-dimensional Planck length and Newton constant by

$$l_p^8 = g^2 \ell_s^8 = G \qquad (12.4.11)$$

On the other side of the correspondence, the gauge theory also has two constants. They are

1) The rank of the gauge group N
2) The gauge coupling g_{ym}

Obviously the two bulk parameters R and g must be determined by N and g_{ym}. In these lectures we will assume the relation that was originally derived in [7].

$$\frac{R}{\ell_s} = (N g_{ym}^2)^{\frac{1}{4}}$$
$$g = g_{ym}^2 \qquad (12.4.12)$$

We can also write the 10-dimensional Newton constant in the form

$$G = R^8/N^2 \qquad (12.4.13)$$

There are two distinct limits that are especially interesting, depending on one's motivation. The AdS/CFT correspondence has been widely studied as a tool for learning about the behavior of gauge theories in the

strongly coupled 't Hooft limit. From the gauge theory point of view the
't Hooft limit is defined by

$$g_{ym} \to 0$$
$$N \to \infty$$
$$g_{ym}^2 N = constant \qquad (12.4.14)$$

From the bulk string point of view the limit is

$$g \to 0$$
$$\frac{R}{\ell_s} = constant \qquad (12.4.15)$$

Thus the strongly coupled 't Hooft limit is also the classical string theory
limit in a fixed and large AdS space. This limit is dominated by classical
supergravity theory.

The interesting limit from the viewpoint of the holographic principle is
a different one. We will be interested in the behavior of the theory as the
AdS radius increases but with the parameters that govern the microscopic
physics in the bulk kept fixed. This means we want the limit

$$g = constant$$
$$R/\ell_s \to \infty \qquad (12.4.16)$$

On the gauge theory side this is

$$g_{ym} = constant$$
$$N \to \infty \qquad (12.4.17)$$

Our goal will be to show that the number of quantum degrees of freedom in
the gauge theory description satisfies the holographic behavior in equation
12.3.8.

12.5   The Infrared Ultraviolet Connection

In either of the metrics in equation 12.2.1 or 12.2.2 the proper area of any
finite coordinate patch tends to ∞ as the boundary of AdS is approached.
Thus we expect that the number of degrees of freedom associated with
such a patch should diverge. This is consistent with the fact that a continuum quantum field theory such as SYM has an infinity of modes in any
finite three-dimensional patch. In order to do a more refined counting[9]

we need to regulate both the area of the AdS boundary and the number of ultraviolet degrees of freedom in the SYM. As we will see, these apparently different regulators are really two sides of the same coin. We have already discussed infrared (IR) regulating the area of AdS by introducing a surrogate boundary L at $r = 1 - \delta$ or similarly at $y = \delta$.

That the IR regulator of the bulk theory is equivalent to an ultraviolet (UV) regulator in the SYM theory is called the IR/UV connection[9]. It is in many ways similar to the behavior of strings as we study them at progressively shorter time scales. In Chapter 14 we will find the interesting behavior that a string appears to grow as we average its properties over smaller and smaller time scales. To understand the relation between this phenomenon and the IR/UV connection in AdS we need to discuss the relation between $AdS(5) \otimes S(5)$ and D-branes.

D-branes are objects which occur in superstring theory. They are stable "impurities" of various dimensionality that can appear in the vacuum. A Dp-brane is a p-dimensional object. We are especially interested in D3-branes. Such objects fill 3 dimensions of space and also time. Their properties are widely studied in string theory and we will only quote the results that we need. The most important property of D3-branes is that they are embedded in a 10-dimensional space. Let us assume that they fill time and the 3 spatial coordinates x^i. Let the other 6 coordinates be called z^m and let $z \equiv \sqrt{z^m z^m}$. We will place a "stack" of N D3-branes at $z = 0$.

Now a single D-brane has local degrees of freedom. For example the location in z may fluctuate. Thus we can think of the z location as a scalar field living on the D-brane. In addition there are modes of the brane which are described by vector fields with components in the t, x direction as well as fermionic modes needed for supersymmetry. Our main concern will be with the $z(x, t)$ fluctuations whose action is known from string theory calculations to be a that of conventional 3+1 dimensional scalar field theory.

D-branes can also be juxtaposed to form stacks of D-branes. A stack of N D-branes has a mass and D-brane charge which grow with N. The mass and charge are sources of bulk fields such as the gravitational field. What makes the D-brane stack interesting to us is that the geometry sourced by the stack is exactly that of $AdS(5) \otimes S(5)$. In fact the geometry defined by 12.2.3 and 12.2.4 is closely related to that of a D-brane stack.

Specifically the geometry sourced by the D-branes is a particular solution of the supergravity equations of motion:

$$ ds^2 = F(z)(dt^2 - dx^2) - F(z)^{-1}dzdz \qquad (12.5.18) $$

where

$$F(z) = \left(1 + \frac{cg_sN}{z^4}\right)^{-1/2} \tag{12.5.19}$$

and c is a numerical constant. If we consider the limit in which $\frac{cg_sN}{z^4} \gg 1$ then we can replace $F(z)$ by the simpler expression

$$F(z) \cong \frac{z^2}{(cg_sN)^{\frac{1}{2}}}. \tag{12.5.20}$$

It is then a simple exercise to see that the D-brane metric is of the form 12.2.3, 12.2.4.

Furthermore the theory of the fluctuations of the stack is $\mathcal{N} = 4$ SYM. All of the fields in this theory form a single supermultiplet belonging to the adjoint ($N \times N$ matrix) representation of $SU(N)$.

In this lecture we give an argument for the IR/UV connection based on the quantum fluctuations of the positions of the D3-branes which are nominally located at the origin of the coordinate z in equation 12.2.3. The location of a point on a 3-brane is defined by six coordinates z, w_5. We may also choose the six coordinates to be Cartesian coordinates $(z^1, ..., z^6)$. The original coordinate z is defined by

$$z^2 = (z^1)^2 + ... + (z^6)^2 \tag{12.5.21}$$

As we indicated, the coordinates z^m are represented in the SYM theory by six scalar fields on the world volume of the branes. If the six scalar fields ϕ^n are canonically normalized, then the precise connection between the z's and ϕ's is

$$z = \frac{g_{ym}\ell_s^2}{R^2}\phi \tag{12.5.22}$$

Strictly speaking equation 12.5.22 does not make sense because the fields ϕ in $SU(N)$ are $N \times N$ matrices, where we identify the N eigenvalues of the matrices in equation 12.5.18 to be the coordinates z^m of the N D3-branes[10]. The geometry is noncommutative and only configurations in which the six matrix valued fields commute have a classical interpretation. However the radial coordinate $z = \sqrt{z^m z^m}$ can be defined by

$$z^2 = \left(\frac{g_{ym}\ell_s^2}{R^2}\right)^2 \frac{1}{N} Tr\phi^2 \tag{12.5.23}$$

A question which is often asked is: where are the D3-branes located in the AdS space? The usual answer is that they are at the horizon $z = 0$. However our experiences in Chapter 14 with similar questions will warn us that the answer may be more subtle. What we will find there is that the way information is localized in space depends on what frequency range it is probed with. High frequency or short time probes see the string widely spread in space while low frequency probes see a well localized string.

To answer the corresponding question about D3-branes we need to study the quantum fluctuations of their position. The fields ϕ are scalar quantum fields whose scaling dimensions are known to be exactly $(length)^{-1}$. From this it follows that any of the N^2 components of ϕ satisfies

$$\langle \phi_{ab}^2 \rangle \sim \delta^{-2} \qquad (12.5.24)$$

where δ is the ultraviolet regulator of the field theory. It follows from equation 12.5.20 that the average value of z satisfies

$$< z >^2 \sim \left(\frac{g_{ym} \ell_s^2}{R^2} \right)^2 \frac{N}{\delta^2} \qquad (12.5.25)$$

or, using equation 12.4.12

$$< z >^2 \sim \delta^{-2} \qquad (12.5.26)$$

In terms of the coordinate y which vanishes at the boundary of AdS

$$< y >^2 \sim \delta^2. \qquad (12.5.27)$$

Here it is seen that the location of the brane is given by the ultraviolet cutoff of the field theory on the boundary. Evidently low frequency probes see the branes at $z = 0$ but as the frequency of the probe increases the brane appears to move toward the boundary at $z = \infty$. The precise connection between the UV SYM cutoff and the bulk theory IR cutoff is given by equation 12.5.23.

12.6   Counting Degrees of Freedom

Let us now turn to the problem of counting the number of degrees of freedom needed to describe the region $y > \delta$ [9]. The UV/IR connection implies that this region can be described in terms of an ultraviolet regulated theory with a cutoff length δ. Consider a patch of the boundary with unit coordinate area. Within that patch there are $1/\delta^3$ cutoff cells of size δ. Within each such cell the fields are constant in a cutoff theory. Thus each cell has of order N^2 degrees of freedom corresponding to the $N \otimes N$ components of the adjoint representation of $U(N)$. Thus the number of degrees of freedom on the unit area is

$$N_{dof} \approx \frac{N^2}{\delta^3} \qquad (12.6.28)$$

On the other hand the 8-dimensional area of the regulated patch is

$$A = \frac{R^3}{\delta^3} \times R^5 = \frac{R^8}{\delta^3} \qquad (12.6.29)$$

and the number of degrees of freedom per unit area is

$$\frac{N_{dof}}{A} \sim \frac{N^2}{R^8} \qquad (12.6.30)$$

Finally we may use equation 12.4.13

$$\frac{N_{dof}}{A} \sim \frac{1}{G} \qquad (12.6.31)$$

This is very gratifying because it is exactly what is required by the Holographic Principle.

Chapter 13

Black Holes in a Box

The apparently irreconcilable demands of black hole thermodynamics and the principles of quantum mechanics have led us to a very strange view of the world as a hologram. Now we will return, full circle, to see how the holographic description of $AdS(5) \otimes S(5)$ provides a description of black holes.

We have treated Schwarzschild black holes as if they were states of thermal equilibrium, but of course they are not. They are long-lived objects, but eventually they evaporate. We can try to prevent their evaporation by placing them in a thermal heat bath at their Hawking temperature but that does not work. The reason is that their specific heat is negative; their temperature decreases as their energy or mass increases. Any object with this property is thermodynamically unstable. To see this, suppose a fluctuation occurs in which the black hole absorbs an extra bit of energy from the surrounding heat bath. For an ordinary system with positive specific heat this will raise its temperature which in turn will cause it to radiate back into the environment. The fluctuations are self-regulating. But a system with negative specific heat will lower its temperature when it absorbs energy and will become cooler than the bath. This in turn will favor an additional flow of energy from the bath to the black hole and a runaway will occur. The black hole will grow indefinitely. If on the other hand the black hole gives up some energy to the environment it will become hotter than the bath. Again a runaway will occur that leads the black hole to disappear.

A well known way to stabilize the black hole is to put it in a box so that the environmental heat bath is finite. When the black hole absorbs some energy it cools but so does the finite heat bath. If the box is not too big the heat bath will cool more than the black hole and the flow of heat will be back to the bath. In this lecture we will consider the properties of black holes which are stabilized by the natural box provided by Anti de Sitter space. More specifically we consider large black holes in $AdS(5) \otimes S(5)$ and their holographic description in terms of the $\mathcal{N} = 4$ Yang–Mills theory.

The black holes which are stable have Schwarzschild radii as large or larger than the radius of curvature R. They homogeneously fill the 5-sphere and are solutions of the dimensionally reduced 5-dimensional Einstein equations with a negative cosmological constant. The thermodynamics can be derived from the black hole solutions by first computing the area of the horizon and then using the Bekenstein–Hawking formula.

Before writing the AdS–Schwarzschild metric, let us write the metric of AdS in a form which is convenient for generalization.

$$d\tau^2 = \left(1 + \frac{r^2}{R^2}\right) dt^2 - \left(1 + \frac{r^2}{R^2}\right)^{-1} dr^2 - r^2 d\Omega^2 \tag{13.0.1}$$

where in this formula r runs from 0 to the boundary at $r = \infty$. Note that the coordinates r, t are not the same as in equation 12.2.1.

The AdS black hole is given by modifying the function $\left(1 + \frac{r^2}{R^2}\right)$:

$$d\tau^2 = \left(1 + \frac{r^2}{R^2} - \frac{\mu G}{R^5 r^2}\right) dt^2 - \left(1 + \frac{r^2}{R^2} - \frac{\mu G}{R^5 r^2}\right)^{-1} dr^2 - r^2 d\Omega^2 \tag{13.0.2}$$

where the parameter μ is proportional to the mass of the black hole and G is the 10-dimensional Newton constant. The horizon of the black hole is at the largest root of

$$\left(1 + \frac{r^2}{R^2} - \frac{\mu G}{R^5 r^2}\right) = 0$$

The Penrose diagram of the AdS black hole is shown in Figure 13.1. One finds that the entropy is related to the mass by

$$S = c \left(M^3 R^{11} G^{-1}\right)^{\frac{1}{4}} \tag{13.0.3}$$

where c is a numerical constant. Using the thermodynamic relation $dM = TdS$ we can compute the relation between mass and temperature:

$$M = c\frac{R^{11} T^4}{G} \tag{13.0.4}$$

Fig. 13.1 *Penrose diagram of the AdS black hole*

or in terms of dimensionless SYM quantities

$$E_{sym} = c\frac{R^8}{G}T_{sym}^4$$

$$= cN^2T_{sym}^4 \qquad (13.0.5)$$

Equation 13.0.5 has a surprisingly simple interpretation. Recall that in $3 + 1$ dimensions the Stephan–Boltzmann law for the energy density of radiation is

$$E \sim T^4V \qquad (13.0.6)$$

where V is the volume. In the present case the relevant volume is the dimensionless 3-area of the unit boundary sphere. Furthermore there are $\sim N^2$ quantum fields in the $U(N)$ gauge theory so that apart from a numerical constant equation 13.0.5 is nothing but the Stephan–Boltzmann law for black body radiation. Evidently the holographic description of the AdS black holes is as simple as it could be; a black body thermal gas of N^2 species of quanta propagating on the boundary hologram.

The constant c in equation 13.0.3 can be computed in two ways. The first is from the black hole solution and the Bekenstein–Hawking formula. The second way is to calculate it from the boundary quantum field theory in the free field approximation. The calculations agree in order of magnitude, but the free field gives too big a coefficient by a factor of 4/3. This is not too surprising because the classical gravity approximation is only valid when g_{YM}^2N is large.

13.1   The Horizon

The high frequency quantum fluctuation of the location of the D3-branes are invisible to a low frequency probe. Roughly speaking this is insured by the renormalization group as applied to the SYM description of the branes. The renormalization group is what insures that our bodies are not severely damaged by constant exposure to high frequency vacuum fluctuations. We are not protected in the same way from classical fluctuations. An example is the thermal fluctuations of fields at high temperature. All probes sense thermal fluctuations of the brane locations. Let us return to equation 12.5.24 but now, instead of using equation 12.5.25 we use the thermal field fluctuations of ϕ. For each of the N^2 components the thermal fluctuations have the form

$$< \phi^2 >= T_{sym}^2 \qquad (13.1.7)$$

and we find equations 12.5.26 and 12.5.27 replaced by

$$\begin{aligned} < z >^2 &\sim T_{sym}^2 \\ < y >^2 &\sim T_{sym}^{-2} \end{aligned} \qquad (13.1.8)$$

It is clear that the thermal fluctuations will be strongly felt out to a coordinate distance $z = T_{sym}$. In terms of r the corresponding position is

$$1 - r \sim 1/T_{sym} \qquad (13.1.9)$$

In fact this coincides with the location of the horizon of the AdS black hole.

13.2   Information and the AdS Black Hole

The AdS black hole is an ideal laboratory for investigating how bulk quantum field theory fails when applied to the fine details of Hawking radiation. Let us consider some field that appears in the supergravity description of the bulk. Such objects are 10-dimensional fields and should not be confused with the 4-dimensional quantum fields associated with the boundary theory. A simple example is the minimally coupled scalar dilaton field ϕ. We will only consider dilaton fields which are constant on the 5-sphere. In that case the action for ϕ is the minimally coupled scalar action in 5-dimensional AdS.

$$I = \int d^5x \sqrt{-g} g^{\mu\nu} \partial_\mu \phi \partial_\nu \phi. \tag{13.2.10}$$

The appropriate boundary conditions are $\phi \to 0$ at the boundary of the AdS, i.e. $r \to \infty$.

Plugging in the black hole metric given in equation 13.0.2, we find a number of things. First $\phi(r) \sim r^{-4}\Phi$ as $r \to \infty$. It is the value of Φ on the boundary which is identified with a local field in the boundary Super Yang–Mills theory. This is true both in pure AdS as well as in the AdS–Schwarzschild metrics. Secondly, in the pure AdS background ϕ is periodic in time, but in the black hole metric ϕ goes to zero exponentially with time:

$$\phi \to exp{-\gamma t}. \tag{13.2.11}$$

Equation 13.2.11 has implications for quantum correlation functions in the black hole background. Consider the correlator $\langle \phi(t)\phi(t') \rangle$. Equation 13.2.11 requires it to behave like

$$\langle \phi(t)\phi(t') \rangle \sim exp{-\gamma |t - t'|}. \tag{13.2.12}$$

for large $|t - t'|$ The parameter γ depends on the black hole mass or temperature and has the form

$$\gamma = \frac{H(\mu G R^{-7})}{R} \tag{13.2.13}$$

where H is a dimensionless increasing function of its argument.

The meaning of this exponential decrease of the correlation function is that the effects of an initial perturbation at time t dissipate away and are eventually lost. In other words the system does not preserve any memory of the initial perturbation. This type of behavior is characteristic of large thermal systems where γ would correspond to some dissipation coefficient. However, exponential decay is not what is really expected for systems of finite entropy such as the AdS black hole that we are dealing with. Any quantum system with finite entropy preserves some memory of a perturbation. Since AdS is exactly described by a conventional quantum system it follows that the correlator should not go to zero. We shall now prove this assertion.

The essential point is that any quantum system with finite thermal entropy must have a discrete spectrum. This is because the entropy is essentially the logarithm of the number of states per unit energy. Indeed the spectrum of the boundary quantum field theory is obviously discrete since it is a theory defined on a finite sphere.

Let us now consider a general finite closed system described by a thermal density matrix and a thermal correlator of the form

$$F(t) = \langle A(0)A(t) \rangle = \frac{1}{Z} Tr e^{-\beta H} A(0) e^{iHt} A(0) e^{-iHt}. \qquad (13.2.14)$$

By finite we simply mean that the spectrum is discrete and the entropy finite. Inserting a complete set of (discrete) energy eigenstates gives

$$F(t) = \frac{1}{Z} \sum_{ij} e^{-\beta E_i} e^{i(E_j - E_i)t} |A_{ij}|^2. \qquad (13.2.15)$$

For simplicity we will assume that the operator A has no matrix elements connecting states of equal energy. This means that the time average of F vanishes.

Let us now consider the long time average of $F(t)F^*(t)$.

$$L = \lim_{T \to \infty} \frac{1}{2T} \int_{-T}^{+T} dt F(t) F^*(t) \qquad (13.2.16)$$

Using equation 13.2.15 it is easy to show that the long time average is

$$L = \frac{1}{Z^2} \sum_{ijkl} e^{-\beta(E_i + E_k)} |A_{ij}|^2 |A_{kl}|^2 \delta_{(E_j - E_l + E_k - E_i)}. \qquad (13.2.17)$$

where the delta function is defined to be zero if the argument is nonzero and 1 if it is zero. The long time average L is obviously nonzero and positive. Thus it is not possible for the correlator $F(t)$ to tend to zero as the time tends to infinity and the limits required by the AdS/CFT correspondence cannot exist. The value of the long time average for such finite systems can be estimated, and it is typically of the order e^{-S} where S is the entropy of the system. This observation allows us to understand why it tends to zero in the (bulk) QFT approximation. In studying QFT in the vicinity of a horizon we have seen that the entropy is UV divergent. This is due to the enormous number of short wave length modes near the horizon. This leads us to a very important and general conclusion: any phenomenon which crucially depends on the finiteness of horizon entropy will be gotten wrong by the approximation of QFT in a fixed background. This includes questions of information loss and of particular interest in this lecture, the long time behavior of correlation functions.

How exactly do the correlations behave in the long time limit? The answer is not that they uniformly approach constants given by the long time averages. The expected behavior is that they fluctuate chaotically. A large fluctuation which reduces the entropy by amount ΔS has probability $e^{-\Delta S}$. Thus we can expect large fluctuations in the correlators at intervals of order e^S. These fluctuations are analogous to the classical phenomenon of Poincaré recurrences. It is generally found that the large time behavior of correlators is chaotic "noise" with the long time average given by

$$L \sim e^{-S}. \tag{13.2.18}$$

This long time behavior, missed by bulk quantum field theory, is a small part of the encoding of information in the thermal atmosphere of the AdS black hole.

PART 3
Black Holes and Strings

Chapter 14

Strings

We have learned a great deal about black holes by considering the behavior of quantum fields near horizons. But ultimately local quantum field theory fails in a number of ways. In general the failures can be attributed to a common cause – quantum field theory has too many degrees of freedom.

The earliest evidence that QFT is too rich in degrees of freedom was the uncontrollable short distance divergences in gravitational perturbation theory. As a quantum field theory, Einstein's general relativity is very badly behaved in the ultraviolet.

Even more relevant for our purposes is the divergence in the entropy per unit area of horizons that was found in Chapter 4. Entropy is a direct measure of the number of active degrees of freedom of a system. Evidently there are far too many degrees of freedom very close to a horizon in QFT. Later in Chapter 12 we quantified just how over-rich QFT is.

The remaining portion of this book deals, in an elementary way, with a theory that seems to have just the right number of degrees of freedom: string theory. The problems posed by black holes for a fundamental theory of quantum gravity are non-perturbative. Until relatively recently, string theory was mostly defined by a set of perturbation rules. Nevertheless, even in perturbation theory, we will see certain trends that are more consistent with black hole complementarity than the corresponding trends in QFT.

In Chapter 9 we explained that the key to understanding black hole complementarity lies in the ultrahigh frequency oscillations of fluctuations of matter in its own rest frame. The extreme red shift between the freely falling frame and the Schwarzschild frame may take phenomena which are of too high frequency to be visible ordinarily and make them visible to the outside observer. As an example, imagine a freely falling whistle that emits a sound of such high frequency that it cannot be heard by the human ear.

As the whistle approaches the horizon, the observer outside the black hole hears the frequency red shifted. Eventually it becomes audible, no matter how high the frequency in the whistle's rest frame.

On the other hand, the freely falling observer who accompanies the whistle never gets the benefit of the increasing red shift. She never hears the whistle.

This suggests that the consistency of black hole complementarity is a deep constraint on how matter behaves at very short times or high frequencies. Quantum field theory gets it wrong, but string theory seems to do better. The qualitative behavior of strings is the subject of this lecture.

In order to compare string theory and quantum field theory near a horizon, we will first study the case of a free particle falling through a Rindler horizon. As we will see, it is natural to use *light cone coordinates* for this problem. The process and conventions are illustrated in Figure 14.1. The coordinates $X^{\pm}$ are defined by

$$X^{\pm} = \frac{X^0 \pm X}{\sqrt{2}} = \mp \frac{\rho}{\sqrt{2}} e^{\mp\tau} \qquad (14.0.1)$$

and the metric is given by

$$d\tau^2 = 2dX^+ dX^- - \left(dX^i\right)^2 \qquad (14.0.2)$$

where X^i run over the coordinates in the plane of the horizon. We will refer to X^i as the transverse coordinates, because they are transverse to the direction of motion of the point particle. The trajectory of the particle is taken to be

$$X^i = 0$$
$$\qquad (14.0.3)$$
$$X^- - X^+ = \sqrt{2}L$$

As the particle falls closer and closer to the horizon, the constant τ surfaces become more and more light-like in the particle's rest frame. In other words, the particle and the Rindler observer are **boosted** relative to one another by an ever-increasing boost angle.

Near the particle trajectory X^+ and τ are related by

$$X^+ \cong -2L\,e^{-2\tau} \qquad (14.0.4)$$

for large τ. This suggests that the description of mechanics in terms of the Rindler (or Schwarzschild) time be replaced by a description in light cone

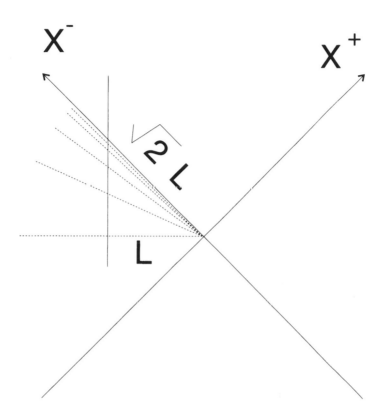

Fig. 14.1   *Free particle falling through a Rindler horizon*

coordinates with X^+ playing the role of the independent time coordinate. We will therefore briefly review particle mechanics in the light cone frame.

14.1   Light Cone Quantum Mechanics

In order to write the action for a relativistic point particle we introduce a parameter σ along the world line of the particle. Since the action only depends on the world line and not the way we parameterize it, the action should be invariant under a reparameterization. Toward this end we also introduce an "einbein" $e(\sigma)$ that transforms under σ-reparameterizations.

$$e^1(\sigma^1)\,d\sigma^1 \;=\; e(\sigma)\,d\sigma \tag{14.1.5}$$

The action is given by

$$W \;=\; \int L\,d\sigma$$

$$L \;=\; -\tfrac{1}{2}\left[\frac{1}{e}\frac{dX^\mu}{d\sigma}\frac{dX_\mu}{d\sigma} - e\,m^2\right] \tag{14.1.6}$$

where m is the mass of the particle. The action in equation 14.1.6 is invariant under σ-reparameterizations.

Let us now write equation 14.1.6 in terms of light cone coordinates and, at the same time use our gauge freedom to fix $\sigma = X^+$ (which is then treated as a time variable). Then L takes the form

$$L \;=\; \frac{1}{2}\left[-\frac{2}{e}\frac{dX^-}{d\sigma} + \frac{1}{e}\frac{dX^i}{d\sigma}\frac{dX^i}{d\sigma} - e\,m^2\right] \tag{14.1.7}$$

The conserved canonical momenta are given by

$$P_- \;=\; \frac{\partial L}{\partial \dot X^-} \;=\; -\frac{1}{e}$$

$$P_i \;=\; \frac{\partial L}{\partial \dot X^i} \;=\; \frac{1}{e}\,\dot X^i \tag{14.1.8}$$

where dot refers to σ derivative. Note that the conservation of P_- insures that $e(\sigma)$ has a fixed constant value.

The Hamiltonian is easily obtained by the standard procedure:

$$H \;=\; \frac{eP_i^2}{2} + \frac{m^2 e}{2} \tag{14.1.9}$$

This form of H manifests a well known fact about light cone physics. If we focus on the transverse degrees of freedom, the Hamiltonian has all the properties of a non-relativistic system with Galilean symmetry. The second term in H is just a constant, and can be interpreted as an internal energy that has no effect on the transverse motion. The first term has the usual non-relativistic form with e^{-1} playing the role of an effective transverse mass. This Hamiltonian and its associated quantum mechanics exactly describes the point particles of conventional free quantum field theory formulated in the light cone gauge.

Now let us consider the transverse location of the particle as it falls toward the horizon. In particular, suppose the particle is probed by an experiment which takes place over a short time interval δ just before horizon crossing. In other words, the particle is probed over the time interval

$$-\delta < X^+ < 0 \tag{14.1.10}$$

by a quantum of (Minkowski) energy $\sim \frac{1}{\delta}$. This experiment is similar to the one discussed in Chapter 9.3, except that the probe carries out information about the transverse location of the particle instead of its baryon number.

Since the interaction is spread over the time interval in equation 14.1.10, the instantaneous transverse position should be replaced by the time averaged coordinate X^i_δ

$$X^i_\delta = \frac{1}{\delta} \int_{-\delta}^{0} X^i(\sigma)\, d\sigma \tag{14.1.11}$$

To evaluate equation 14.1.11, we use the non-relativistic equations of motion

$$X^i(\sigma) = X^i(0) + eP^i\sigma \tag{14.1.12}$$

to give

$$X^i_\delta = X^i(0) + \frac{eP^i\delta}{2}. \tag{14.1.13}$$

Finally, let us suppose that the particle wave function is initially a smooth wave packet well localized in transverse position with uncertainty ΔX^i. Let us also assume the very high momentum components of the wave function are negligible. Under these conditions nothing singular happens to the probability distribution for X^i_δ as $\delta \to 0$. No matter how small δ is, the effective probability distribution for X_δ is concentrated in a well localized region of fixed extent, δX. There is no tendency for information to transversely spread over a stretched horizon.

All of this is exactly what is expected for an ordinary particle in free quantum field theory. For the more interesting case of an interacting quantum field theory, we could study the transverse properties of an interacting or composite particle such as a hydrogen atom. For example, a time averaged relative coordinate or charge density can be defined, and it too shows no sign of spreading as the sampling interval δ tends to zero.

Why is this a problem? The reason is that it conflicts with the complementarity principle. Complementarity requires the probe to report that the particle fell into a very high temperature environment in which it repeatedly suffered high energy collisions. In this kind of environment the information stored in the infalling system would be thermalized and spread over the horizon. The implication for the probing experiment is that the particle should somehow spread or diffuse roughly the way the effective charge distribution did in Chapter 7.

14.2   Light Cone String Theory

Although naive pertubative string theory cannot capture this effect completely correctly, the tendency is already there in the theory of free strings. A free string is a generalization of a free particle. There are a number of excellent textbooks on string theory that the reader who is interested in technical details can consult. For our purposes, only the most elementary aspects of string theory will be needed.

A string is a one-dimensional continuum whose points are parameterized by a continuous parameter σ^1. The transverse coordinates of the point at σ^1 are labeled $X^i(\sigma)$, where σ^1 runs from 0 to 2π. It is also a function of a time-like parameter σ^0, which is identified with light cone time X^+. Thus $X^i(\sigma^0, \sigma^1)$ is a field defined on a 1+1 dimensional parameter space (σ^a). In addition to $X^i(\sigma)$, the canonical momentum density $P_i(\sigma)$ can also be defined. At equal times X and P satisfy

$$\left[X^i(\sigma), P_j(\sigma')\right] = i\,\delta^i_j\,\delta(\sigma - \sigma') \tag{14.2.14}$$

The light cone Hamiltonian for the free string is a natural generalization of that for a free particle;

$$H = \frac{1}{P_-} \int_0^{2\pi} \frac{d\sigma'}{2} \left[|P_i(\sigma')|^2 + \left(\frac{\partial X^i}{\partial \sigma'}\right)^2\right] \tag{14.2.15}$$

We have used units in which the string tension (energy per unit length in the rest frame) is unity.

The equation of motion following from equations 14.2.14 and 14.2.15 is a simple wave equation

$$\frac{\partial^2 X^i}{(\partial \sigma^0)^2} - \frac{\partial^2 X^i}{(\partial \sigma^1)^2} = 0. \tag{14.2.16}$$

Quantization of the string is straightforward. $X^i(\sigma)$ becomes a free scalar field in 1+1 dimensions satisfying equation 14.2.16 with periodic boundary conditions in σ^1, $X(\sigma^0, 2\pi) = X(\sigma^0, 0)$.

The string differs in important ways from the free particle, especially in its short time behavior. As we have repeatedly emphasized, it is the short time behavior that is key to complementarity.

Let us consider the analog of the question that we addressed about the time averaged location of the point particle. Now we consider the time averaged location of a point on the string. Thus, define

$$X_\delta = \frac{1}{\delta} \int_0^\delta d\sigma^0 X(\sigma) \qquad (14.2.17)$$

Since all points σ^1 are equivalent, it doesn't matter what value σ^1 takes on the right hand side when we evaluate X_δ. A useful measure of how much the information in a string is spread as it falls towards the horizon is provided by the fluctuations in X_δ, that is

$$(\Delta X)^2 = \langle X_\delta^2 \rangle - \langle X_\delta \rangle^2 \qquad (14.2.18)$$

The state used for the expectation value in equation 14.2.18 is the ground state string. This quantity is easily calculated and diverges logarithmically as $\delta \to 0$. In other words, as the string approaches the horizon, any experiment (from the outside) to determine how its internal parts are distributed will indicate a logarithmic increase in the area it occupies

$$(\Delta X)^2 \sim |log\,\delta|. \qquad (14.2.19)$$

Another way to write equation 14.2.19 is to use the connection between Rindler time and light cone time in equation 14.0.4

$$(\Delta X)^2 \sim |log\,(2Le^{-2\tau})| \sim 2\tau.$$

Finally, we can use the relation betwen Rindler time and Schwarzschild time given by $\tau = t/4MG$ to obtain

$$(\Delta X)^2 \sim \frac{\alpha' t}{4MG}. \qquad (14.2.20)$$

In equation 14.2.20 we have restored the units by including the factor α', the inverse string tension.

Here we see the beginnings of an explanation of complementarity. The observer outside the black hole will find the string diffusing over an increasing area of the horizon as time progresses. But an observer falling with the string and doing low energy experiments on it would conclude that the string remains a fixed finite size as it falls.

The linear growth of the area in equation 14.2.20 is much slower than the growth of a charged particle described in Chapter 7. In that case inspection of 7.0.21 indicates that the growth is exponential. A completely consistent theory would require these growth patterns to match. The true exponential asymptotic growth is undoubtedly a non-perturbative phenomenon that involves string interactions in an essential way.

To see how interactions influence the evolution, let's determine the average total length of string, projected onto the two-dimensional transverse plane

$$\ell = \int_0^{2\pi} d\sigma^1 \sqrt{\left|\frac{\partial X^i}{\partial \sigma^1}\right|^2} \tag{14.2.21}$$

As a preliminary, let us consider the ground state average of $\left|\frac{\partial X^i}{\partial \sigma^1}\right|^2$. This is another exercise in free scalar quantum field theory, and the result is quadratically divergent.

If however $\frac{\partial X^i}{\partial \sigma}$ is averaged over the time interval δ, we find that the ground state average of $\left|\frac{\partial X^i}{\partial \sigma^1}\right|^2$ is given by

$$\left\langle \frac{\partial X^i}{\partial \sigma^1} \frac{\partial X^i}{\partial \sigma^1} \right\rangle \sim \frac{1}{\delta^2} \tag{14.2.22}$$

Using the fact that the probability distribution for $\frac{\partial X^i}{\partial \sigma^1}$ is Gaussian in free field theory, we can conclude that $\left\langle \sqrt{\left|\frac{\partial X}{\partial \sigma^1}\right|^2} \right\rangle$ or ℓ scales as

$$\ell \approx \frac{1}{\delta} \tag{14.2.23}$$

or using equation 14.0.4

$$\ell \approx \frac{1}{2L} e^{2\tau}. \tag{14.2.24}$$

In other words, as the string falls toward the horizon, it grows exponentially in length.

Another quantity which exponentially grows is the ρ component of the Rindler momentum. To see this, we use the transformation in equation 14.0.1 to derive

$$\frac{\partial}{\partial \rho} = \frac{1}{\sqrt{2}} \left[-e^{-\tau} \frac{\partial}{\partial X^+} - + e^{\tau} \frac{\partial}{\partial X^-} \right],$$

or in terms of momenta

$$P_\rho = \frac{1}{\sqrt{2}} \left(e^\tau P_- - e^{-\tau} P_+\right) \tag{14.2.25}$$

In the Rindler approximation to a black hole horizon, $P_\pm$ are conserved, and therefore as $\tau \to \infty$ the radial momentum P_ρ grows like e^τ. Evidently then the ratio of the string length to its total radial momentum is fixed. As the string falls toward the horizon, its radial momentum increases by the mechanism of its physical length increasing.

14.3   Interactions

In Chapter 7 we saw that a charge falling toward the stretched horizon spreads over an area which grows exponentially. The area occupied by a free string only grows linearly. However, this not the end of the story. The the total length of the string grows exponentially with τ. It is clear that this behavior cannot continue indefinitely. The exponential growth of string length and linear growth of area imply that the transverse density of string increases to the point where string interactions must become important and seriously modify the free string picture. Roughly speaking, when a piece of string gets within a distance of order $\sqrt{\alpha'}$ of another piece, they can interact. The number of such string encounters will obviously increase without bound as $\tau \to \infty$.

String interactions are governed by a dimensionless coupling constant g which determines the amplitude for strings to rearrange when they cross. Obviously the importance of interactions is governed not only by g, but also by the local density of string crossings. Let ρ be the number of such crossings per unit horizon area. When $g^2\rho$ becomes large, interactions can no longer be ignored.

Now, the form $g^2\rho$ is not dimensionless. There is only one dimensional constant in string theory, the inverse string tension α' with units of area. Sometimes α' is replaced by a length $\ell_s = \sqrt{\alpha'}$. The dimensionally correct statement is that string interactions become important when

$$g^2\rho \geq \frac{1}{\ell_s^2} \approx \frac{1}{g^2\alpha'} \tag{14.3.26}$$

This criterion has a profound significance. The quantity $g^2\alpha'$ in string theory also governs the gravitational interaction between masses. It is none other than the gravitational coupling constant (in units with $c = \hbar = 1$). The implication is that interactions become important when the area density of the string approaches $\frac{1}{G}$, the area density of horizon entropy.

Although we cannot follow the string past the point where interactions become important, we can be sure that something new will happen. A good guess is that the density of string saturates at order $\frac{1}{G}$. Since the total length of string grows like $\ell \approx e^{\tau}$ the area that it occupies must also grow exponentially. This is reminiscent of the pattern of growth that we encountered in Chapter 7.

Verifying that as $\delta \to 0$ the string grows as if the density saturates is beyond the current technology of string theory. But the simple assumption that splitting and joining interactions cause effective short range repulsion, and that the repulsion prevents the density from increasing indefinitely, provides a phenomenological description of how information spreads over the horizon. Since the spreading is associated with a decreasing time of averaging it is not seen by a freely falling observer. This is the essence of complementarity.

In general, string theory is not a 4-dimensional theory. It is important to check if the same logic applies in higher dimensions. Let D be the space-time dimension. The general case goes as follows:
Since the number of transverse directions is D-2, equation 14.3.26 is replaced by

$$g^2 \rho \geq \frac{1}{\ell_s^{D-2}} = \frac{1}{(\alpha')^{\frac{D-2}{2}}} \tag{14.3.27}$$

or

$$\rho \approx \frac{1}{g^2 \, \ell_s^{D-2}} \tag{14.3.28}$$

In D dimensions, the gravitational and string couplings are related by

$$G \approx g^2 \, \ell_s^{D-2} \tag{14.3.29}$$

so that the perturbative limit is again reached when the density is of order

$$\rho \approx \frac{1}{G} \tag{14.3.30}$$

14.4 Longitudinal Motion

In discussing the properties of horizons, we have repeatedly run into the idea of the stretched horizon, located a microscopic distance above the mathematical horizon. It is natural to ask if the stretched horizon has any reality or whether it is just a mathematical fiction? Of course, to a Frefo observer, neither the stretched nor the mathematical horizon appears real. But to a Fido the stretched horizon is the layer containing the physical degrees of freedom that give rise to the entropy of the horizon. Thus, consistency requires that in some appropriate sense, the degrees of freedom of an infalling object should get deposited in this layer of finite thickness. To study this question we must examine how strings move in the X^- direction.

It is a curious property of string theory that $X^-(\sigma)$ is not an independent degree of freedom. The only degrees of freedom carried by the string are the transverse coordinates $X^i(\sigma)$ which lie in the plane of the horizon. The longitudinal location is defined by an equation whose origin is in the gauge fixing to the light cone gauge. The derivation is provided in the supplement that follows this discussion.

$$\frac{\partial X^-}{\partial \sigma_1} = \frac{\partial X^i}{\partial \sigma_1} \frac{\partial X^i}{\partial \sigma_0} \qquad (14.4.31)$$

Recall that the quantum theory of $X(\sigma)$ is a simple (1+1 dimensional) quantum field theory of (D-2) free scalar fields defined on a unit circle. In such a theory, local operators can be characterized by a mass dimension. For example, X^i has dimension zero, while $\frac{\partial X^i}{\partial \sigma}$ has dimension 1. The right side of equation 14.4.31 has dimension 2. Thus it is apparent that $X^-(\sigma)$ has dimension 1. It immediately follows that the fluctuation in X^- satisfies

$$\Delta X^- \sim \frac{\ell_s^2}{\delta} \qquad (14.4.32)$$

The factor ℓ_s^2 is needed for dimensional reasons.

Another way to write equation 14.4.32 is to observe that δ is an averaging time in light cone coordinates. In other words $\delta = \Delta X^+$. Equation 14.4.32 then takes the form of an uncertainty principle

$$\Delta X^- \, \Delta X^+ \approx \ell_s^2. \qquad (14.4.33)$$

Now the geometric meaning of equation 14.4.33 is very interesting. Let us draw the motion of the fluctuating string in the $X^\pm$ plane as it falls towards

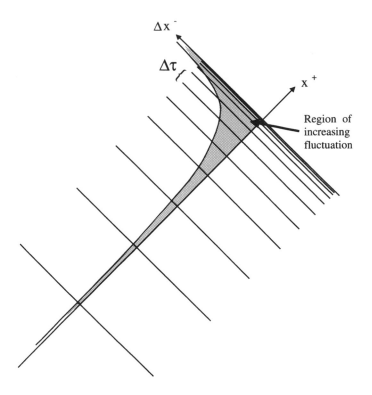

Fig. 14.2   *Near horizon Rindler time slices*

the horizon. Evidently as X^+ tends to zero the fluctuation in X^- required by equation 14.4.33 must increase. This is shown in Figure 14.2. What the figure illustrates is that the stringy material tends to fill a region out to a fixed proper distance from the mathematical horizon at $X^+ = 0$. In other words, unlike a point particle, the stringy substance is seen by a probe to hover at a distance $\sim \ell_s$ above the horizon. Once again this surprising result is a direct consequence of arbitrarily high frequency fluctuations implicit in the stringy structure of matter. Note that if the string coupling satisfies $g_s \ll 1$, the string length ℓ_s can be much greater than the Planck length, $\ell_P = \ell_s g$. In this case it is the string length and not the Planck length that controls the distance of the stretched horizon from the mathematical horizon.

Supplement: Light cone gauge fixing of longitudinal string motions

The coordinate $X^-(\tau,\sigma)$ is not an independent degree of freedom, but is given by equation 14.4.31. To see this multiply 14.4.31 by $1 = \frac{\partial X^+}{\partial \tau}$, which is true for light cone coordinates. Thus we must check the validity of the following equation

$$\frac{\partial X^+}{\partial \tau}\frac{\partial X^-}{\partial \sigma} - \frac{\partial X^j}{\partial \tau}\frac{\partial X^j}{\partial \sigma} = 0. \qquad (14.4.34)$$

The content of this equation expresses the underlying invariance of string theory to a reparameterization of the σ coordinate. Under the transformation

$$\sigma \to \sigma + \delta\sigma$$

the $X's$ transform as

$$X \to X + \frac{\partial X}{\partial \sigma}\delta\sigma. \qquad (14.4.35)$$

The Noether charge for this invariance is exactly the quantity in equation 14.4.34. Setting this quantity to zero insures that the physical spectrum consists only of states which are invariant under σ reparameterization. In going to the light cone frame the constraint serves to define X^- in terms of the transverse coordinates. However even in the light cone gauge there is a bit of residual gauge invariance, namely shifting υ by a constant. In the light cone frame only the transverse $X's$ are dynamical and the generator of these rigid shifts is

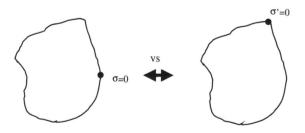

Fig. 14.3 *String parameter translational invariance*

$$\int d\sigma \frac{\partial X^j}{\partial \sigma} \frac{dX^j}{\partial \tau} \tag{14.4.36}$$

which by equation 14.4.31 is equal to

$$\oint d\sigma \frac{\partial X^-}{\partial \sigma}. \tag{14.4.37}$$

Setting this to zero simultaneously insures invariance under shifts of σ and periodicity of $X^-(\sigma)$.

Chapter 15

Entropy of Strings and Black Holes

The Bekenstein–Hawking entropy of black holes points to some kind of microphysical degrees of freedom, but it doesn't tell us what they are. A real theory of quantum gravity should tell us and also allow us to compute the entropy by quantum statistical mechanics, that is, counting microstates. In this lecture we will see to what extent string theory provides the microstructure and to what extent it enables us to compute black hole entropy microscopically.

String theory has many different kinds of black holes, some in $3 + 1$ dimensions, some in higher dimensions. The black holes can be neutral or be charged with the various charges that string theory permits. We will see that for the entire range of such black holes, the statistical mechanics of strings allows us to compute the entropy up to numerical factors of order unity. In every case the results nontrivially agree with the Bekenstein–Hawking formula. What is more, in one or two cases in which the black holes are invariant under a large amount of supersymmetry the calculations can be refined and give the exact numerical coefficients. All of this is in cases where quantum field theory would give an infinite result.

Because string theory is not necessarily a 4-dimensional theory, it is worth exploring the connections between strings and black holes in any dimension. Let us begin with the formula for the entropy of a Schwarzschild black hole in an arbitrary number of dimensions. Call the number of space-time dimensions D. The black hole metric found by solving Einstein's equation in D dimensions is given by

$$d\tau^2 = \left(1 - \frac{R_S^{D-3}}{r^{D-3}}\right) dt^2 - \left(1 - \frac{R_S^{D-3}}{r^{D-3}}\right)^{-1} dr^2 - r^2 \, d\omega_{D-2} \quad (15.0.1)$$

The horizon is defined by

$$R_S = \left(\frac{16\pi(D-3)GM}{\Omega_{D-2}(D-2)} \right)^{\frac{1}{D-3}} \tag{15.0.2}$$

and its D-2 dimensional "area" is given by

$$A = R_S^{D-2} \int d\Omega_{D-2} = R_S^{D-2} \Omega_{D-2}. \tag{15.0.3}$$

Finally, the entropy is given by

$$S = \frac{A}{4G} = \frac{(2GM)^{\frac{D-2}{D-3}} \Omega_{D-2}}{4G}. \tag{15.0.4}$$

The entropy in equation 15.0.4 is what is required by black hole thermodynamics.

Supplement: Schwarzschild geometry in $D = d + 1$ dimensions

To extend a static spherically symmetric geometry to $D = d + 1$ dimensions, the metric can be assumed to be of the form

$$
\begin{aligned}
ds^2 = {} & -e^{2\Phi} dt^2 + e^{2\Delta} dr^2 \\
& + r^2 \left(d\theta_1^2 + sin^2\theta_1 d\theta_2^2 + \dots + sin^2\theta_1 \dots sin^2\theta_{d-2} d\theta_{d-1}^2 \right).
\end{aligned} \tag{15.0.5}
$$

Using orthonormal coordinates, the $G_{\hat{t}}^{\hat{t}}$ component of the Einstein tensor can be directly calculated to be of the form

$$G_{\hat{t}}^{\hat{t}} = -\left[(D-2)\Delta' \frac{e^{-2\Delta}}{r} + \frac{(D-2)(D-3)}{2r^2}(1 - e^{-2\Delta}) \right] \tag{15.0.6}$$

From Einstein's equation for ideal pressureless matter, $G_{\hat{t}\hat{t}} = \kappa\rho$. This means

$$G_{\hat{t}}^{\hat{t}} = -\frac{(D-2)}{2r^{D-2}} \frac{d}{dr} \left[r^{D-3}(1 - e^{-2\Delta}) \right] = -\kappa\rho \tag{15.0.7}$$

which can be solved to give

$$(1 - e^{-2\Delta})r^{D-3} = \frac{2\kappa}{D-2} \int_0^r \rho(r')r'^{D-2} dr' = \frac{2\kappa}{(D-2)} \frac{M}{\Omega_{D-2}} \tag{15.0.8}$$

where the solid angle is given by $\Omega_{D-2} = \frac{2\pi^{(D-1)/2}}{\Gamma((D-1)/2)}$. The $G_{\hat{r}}^{\hat{r}}$ component of the Einstein tensor satisfies

$$
\begin{aligned}
G_{\hat{r}}^{\hat{r}} &= -\left[-(D-2)\Phi' \frac{e^{-2\Delta}}{r} + \frac{(D-2)(D-3)}{2r^2}(1 - e^{-2\Delta}) \right] \\
&= -\left[-(D-2)(\Phi' + \Delta') \frac{e^{-2\Delta}}{r} + \kappa\rho \right]
\end{aligned}
\tag{15.0.9}
$$

For pressureless matter in the exterior region ($\rho = 0 = P$), we can immediately conclude that $\Phi = -\Delta$. Defining the Schwarzschild radius $R_S^{D-3} = \frac{2\kappa M}{(D-2)\Omega_{D-2}}$ we obtain the form of the metric

$$
e^{-2\Delta} = 1 - \left(\frac{R_S}{r} \right)^{D-3} = e^{2\Phi}
\tag{15.0.10}
$$

If we write $F(r) \equiv e^{2\Phi}$, a useful shortcut for calculating the solution to Einstein's equation 15.0.7 is to note its equivalence to the Newtonian Poisson equation in the exterior region

$$
\nabla^2 F(r) = -\kappa\rho, \qquad F(r) = 1 + 2\phi_{Newton}.
\tag{15.0.11}
$$

The Hawking temperature can be calculated by determining the dimensional factor between the Rindler time and Schwarzschild time. Near the horizon, the proper distance to the horizon is given by

$$
\rho = \frac{2R_S}{(D-3)} \sqrt{\left(\frac{r}{R_s} \right)^{D-3} - 1}
\tag{15.0.12}
$$

which gives the relation between Rindler time/temperature units and Schwarzschild time/temperature units

$$
d\omega = \frac{(D-3)}{2R_S} dt
\tag{15.0.13}
$$

Thus, the Hawking temperature of the black hole is given by

$$
T_{Hawking} = \frac{1}{2\pi} \frac{(D-3)}{2R_S}.
\tag{15.0.14}
$$

Using the first law of thermodynamics, the entropy can be directly calculated to be of the form

$$
S = \frac{2\pi(D-3)A}{\kappa}
\tag{15.0.15}
$$

Substituting the form $\kappa = 8\pi(D-3)G$ for the gravitational coupling gives the previous results in D-dimensions.

All calculations of entropy in string theory make use of a well known trick of quantum mechanics. The trick consists of identifying some kind of control parameter that can be adiabtically varied. In the process of adiabatic variation, energy levels are neither created nor destroyed. Thus if we can follow the system to a value of the control parameter where the system is tractable we can count the states easily even if the nature of the object changes during the variation. Basically we are using the quantum analog of the method of adiabatic invariants.

The trick in the string theory context is to vary the strength of the string coupling adiabatically until we arrive at a point where the gravitational forces are so weak that the black hole "morphs" into some more tractable object. Thus we begin with a black hole of mass M_o in a theory with string coupling g_o. Adiabatically varying a control parameter like g_o will cause a change in the black hole mass and other internal structural features. But such a variation will not alter its entropy. Entropy is an adiabatic invariant.

Let us imagine decreasing the string coupling g. What happens to the black hole as g tends to zero? The answer is obvious. It must turn into a collection of free strings. String theory has all kinds of non-perturbative objects, branes of various dimensionality such as membranes, D-branes, monopoles, and so on. But only the free strings have finite energy in the limit $g \to 0$. Therefore a neutral black hole must evolve into a collection of free strings. A very massive black hole might evolve into a large number of low mass strings or, at the opposite extreme, a single very highly excited string.

Very highly excited free strings have an enormously rich spectrum. They can be thought of as a mass of tangled string that forms a time-varying random walk in space. Such random walking strings have a large entropy and can be studied statistically.

The entropy of a string of mass m can be calculated by returning to the light cone quantization of the previous lecture. For any eigenstate of the Hamiltonian with vanishing transverse momentum and unit P_- the light cone energy is $m^2/2$.

On the other hand the quantization of the string defines a 1+1 dimensional quantum field theory in which the (D-2) transverse coordinates $X^i(\sigma)$ play the role of free scalar fields. The spatial coordinate of this field theory is σ_1, and it runs from 0 to 2π.

The counting of the states of a free string is best done in the light cone version of the theory that we discussed in the last lecture. In order to describe the highly excited string spectrum, a formal light cone temperature T can be defined. Recall that the free string is described by means of a $1 + 1$ dimensional quantum theory containing $D - 2$ fields X^i.

The entropy and energy of such a quantum field theory can be calculated by standard means. The leading contribution for large energy is (setting $\ell_s = 1$)

$$E = \pi T^2 (D - 2)$$

$$S = 2\pi T (D - 2)$$

(15.0.16)

Using $E = \frac{m^2}{2}$ and eliminating the temperature yields $S = \sqrt{2(D - 2)\pi}\, m$ or, restoring the units

$$S = \sqrt{2(D - 2)\pi}\, m\, \ell_s.$$

(15.0.17)

Subleading corrections can also be calculated to give

$$S = \sqrt{2(D - 2)\pi}\, m\, \ell_s - c\, log\, (m\, \ell_s)$$

(15.0.18)

where c is a positive constant. The entropy is the log of the density of states. Therefore the number of states with mass m is

$$N_m = \left(\frac{1}{m\, \ell_s}\right)^c exp\left(\sqrt{2\pi(D - 2)}\, m\ell_s\right)$$

(15.0.19)

The formula 15.0.19 is correct for the simplest bosonic string, but similar formulae exist for the various versions of superstring theory.

Now let us compare the entropy of the single string with that of n strings, each carrying mass $\frac{m}{n}$. Call this entropy $S_n(m)$. Then

$$S_n(m) = n\, S(m/n)$$

(15.0.20)

or

$$S_n(m) = \sqrt{2(D - 2)\pi}\, m\, \ell_s - n\, c\, log\left(\frac{m\, \ell_s}{n}\right)$$

(15.0.21)

Obviously for large n the single string is favored. This is actually quite general. For a given total mass, the statistically most likely state in free string theory is a single excited string. Thus it is expected that when the string coupling goes to zero, most of the black hole states will evolve into a single excited string.

These observations allow us to estimate the entropy of a black hole. The assumptions are the following:

- A black hole evolves into a single string in the limit $g \to 0$

- Adiabatically sending g to zero is an isentropic process; the entropy of the final string is the same as that of the black hole

- The entropy of a highly excited string of mass m is of order

$$S \sim m\ell_s \qquad (15.0.22)$$

- At some point as $g \to 0$ the black hole will make a transition to a string. The point at which this happens is when the horizon radius is of the order of the string scale.

To understand this last assumption begin with a massive black hole. Gravity is clearly important and cannot be ignored. But no matter how massive the black hole is, as we decrease g a point will come where the gravitational constant is too weak to matter. That is the point where the black hole makes a transition and begins to act like a string.

The string and Planck length scales are related by

$$g^2 \ell_s^{D-2} = \ell_p^{D-2}. \qquad (15.0.23)$$

Evidently as g decreases the string length scale becomes increasingly big in Planck units. Eventually, at some value of the coupling that depends on the mass of the black hole, the string length will exceed the Schwarzschild radius of the black hole. This is the point at which the transition from black hole to string occurs. In what follows we will vary the g while keeping fixed the string length ℓ_s. This implies that the Planck length varies.

Let us begin with a black hole of mass M_o in a string theory with coupling constant g_o. The Schwarzschild radius is of order

$$R_S \sim (M_o G)^{\frac{1}{D-3}}, \qquad (15.0.24)$$

and using

$$G \approx g^2 \ell_s^{D-2} \qquad (15.0.25)$$

we find

$$\frac{R_S}{\ell_s} \approx \left(\ell_s M_o g_o^2\right)^{\frac{1}{D-3}}. \qquad (15.0.26)$$

Thus for fixed g_o if the mass is large enough, the horizon radius will be much bigger than ℓ_s.

Now start to decrease g. In general the mass will vary during an adiabatic process. Let us call the g-dependent mass $M(g)$. Note

$$M(g_o) = M_o \qquad (15.0.27)$$

The entropy of a Schwarzschild black hole (in any dimension) is a function of the dimensionless variable $M\,\ell_P$. Thus, as long as the system remains a black hole,

$$M(g)\,\ell_P = constant. \qquad (15.0.28)$$

Since $\ell_P \approx \ell_s\, q^{\frac{2}{D-2}}$ we can write equation 15.0.28 as

$$M(g) = M_o \left(\frac{g_o^2}{g^2} \right)^{\frac{1}{D-2}}. \qquad (15.0.29)$$

Now as $g \to 0$ the ratio of the g-dependent horizon radius to the string scale decreases. From equation 15.0.2 it becomes of order unity at

$$M(g)\,\ell_P^{D-2} \approx \ell_s^{D-3} \qquad (15.0.30)$$

which can be written

$$M(g)\,\ell_s \approx \frac{1}{g^2}. \qquad (15.0.31)$$

Combining equations 15.0.29 and 15.0.31 we find

$$M(g)\,\ell_s \approx M_o^{\frac{D-2}{D-3}}\, G_o^{\frac{1}{D-3}}. \qquad (15.0.32)$$

As we continue to decrease the coupling, the weakly coupled string mass will not change significantly. Thus we see that a black hole of mass M_o will evolve into a free string satisfying equation 15.0.32. But now we can compute the entropy of the free string. From equation 15.0.22 we find

$$S \approx M_o^{\frac{D-2}{D-3}}\, G_o^{\frac{1}{D-3}}. \qquad (15.0.33)$$

This is a very pleasing result in that it agrees with the Bekenstein–Hawking entropy in equation 15.0.4. However, in this calculation the entropy is calculated as the microscopic entropy of fundamental strings.

The evolution from black hole to string can be pictorially represented by starting with a large black hole. The stretched horizon is composed of a

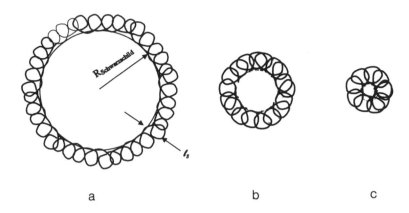

Fig. 15.1 *Evolution from black hole to string. (a) A black hole with stringy stretched horizon smaller than Schwarzschild radius, (b) with stretched horizon and string scale comparable to radius scale, and (c) turned into a string*

stringy mass to a depth of $\rho = \ell_s$ as in the diagram Figure 15.1a. The area density of string is saturated at $\sim \frac{1}{G}$. Another important property of the stretched horizon is its proper temperature. Since the proper temperature of a Rindler horizon is $\frac{1}{2\pi\rho}$, the temperature of the stringy mass will be

$$T_{Stretched} \approx \frac{1}{\ell_s}$$

This temperature is close to the so-called Hagedorn temperature, the maximum temperature that a string can achieve.

As the Schwarzschild radius is decreased (in string units), the area of the horizon decreases but the depth of the stretched horizon stays fixed as in Figure 15.1b. Finally the horizon radius is no larger than ℓ_s (Figure 15.1c) and the black hole turns into a string.

By now a wide variety of black holes that occur in string theory have been analyzed in this manner. The method is always the same. We adiabatically allow g to go to zero and identify the appropriate string configuration that the black hole evolves into.

A particularly interesting situation is that of charged extremal black holes which may be supersymmetric configurations of a supersymmetric theory. In this case the extremal black hole is absolutely stable and in addition, its mass is completely determined by supersymmetry. When this occurs there is no need to follow the mass of the black hole as g varies; the mass is fixed. Under these conditions the black hole can be compared directly to the corresponding weakly coupled string configuration and the entropy read off from the degeneracy of the string theory spectrum. In the cases where exact calculations are possible the charges carried by black holes are associated not with fundamental strings but D-branes. Nevertheless the principles are that same as those that we used to study the Schwarzschild black hole in D dimensions. The results in these more complicated examples are in precise agreement with the Hawking–Bekenstein entropy.

Hagedorn Temperature Supplement

On general grounds, one can determine the density of states η for the various string modes m:

$$\eta(m) \sim exp(4\pi m\sqrt{\alpha'})$$

This allows the partition function to be written as

$$Z \sim \int_0^\infty exp(4\pi m\sqrt{\alpha'})exp\left(-\tfrac{m}{T}\right) dm$$

which diverges if the temperature T is greater than the Hagedorn temperature defined by

$$T_{Hagedorn} \equiv \tfrac{1}{4\pi\sqrt{\alpha'}}$$

The Hagedorn temperature scales with the inverse string length

$$T_{Hagedorn} \sim \tfrac{1}{l_s}.$$

To get a feel for the scale of the Hagedorn temperature, recall the behavior of the entropy given by $S\sim$log(density of states). Using dimensional considerations, we have seen that the entropy of the string scales like

$$S_{string} \sim \sqrt{d_f}\, M_s\, l_s,$$

where d_f is the number of internal degrees of freedom available. Thus, the density of states behaves like

$$e^{S_{string}} \sim e^{\sqrt{d_f}M_s l_s} \sim e^{1/T_{Hagedorn}}$$

which gives the scale $T_{Hagedorn}\sim 1/l_s$. If one examines multi-string fluctuations as a function of temperature, the Hagedorn temperature is the "percolation" temperature for multiple strings fluctuations to coalesce into fluctuations of a single string as represented in Figure 15.2.

Fig. 15.2   *String "percolation"*

Conclusions

The views of space and time that held sway during most of the 20th century were based on locality and field theory, first classical field theory and later quantum field theory. The most fundamental object was the space-time point or better yet, the event. Although quantum mechanics made the event probabilistic and relativity made simultaneity non-absolute, it was assumed that all observers would agree on the usual invariant relationships between events. This view persisted even in classical general relativity. But the paradigm is gradually shifting. It was never adequate to deal with the combination of quantum mechanics and general relativity.

The first sign of this was the failure of standard quantum field theory methods when applied to the Einstein action. For a long time it was assumed that this just meant that the theory was incomplete at short distances in the same way that the Fermi theory of weak interactions was incomplete. But the dilemma of apparent information loss in black hole physics that was uncovered by Hawking in 1976 said otherwise. In order to reconcile the equivalence principle with the rules of quantum mechanics the rules of locality have to be massively modified. The problem is not a pure ultraviolet problem but an unprecedented mix of short distance and long distance physics. Radical changes are called for.

The new paradigm that is gradually emerging is based on four closely related concepts. The first is Black Hole Complementarity. This principle is a new kind of relativity in which the location of phenomena depends on the resolution time available to the experimenter who probes the system. An extreme example would be the fate of someone, call her Alice, falling into an enormous black hole with Schwarzschild radius of a billion years. According to the low frequency observer, namely Alice herself, or someone falling with her, nothing special is felt at the horizon. The horizon is harmless and she or her descendants can live for a billion years before being crushed at the singularity.

In apparent complete contradiction, the high frequency observer who stays outside the black hole finds that his description involves Alice falling into a hellish region of extreme temperature, being thermalized, and eventually re-emitted as Hawking radiation. All of this takes place just outside the mathematical horizon. Obviously this has to do with more than just a modification of the short distance physics. As we have seen, the key to black hole complementarity is the extreme red shift of the quantum fluctuations as seen by the external observer.

The second new idea is the Infrared/Ultraviolet connection. Very closely related to Black Hole Complementarity, the IR/UV connection reverses one of the most fundamental trends of 20th century physics. Throughout that century a close connection between energy and size prevailed. If one wished to study progressively smaller and smaller objects one had to use higher and higher energy probes. But once gravity is involved that trend is reversed. At energies above the Planck scale any possible short distance physics that we might look for is shrouded behind a black hole horizon. As we raise the energy we wind up probing larger and larger distance scales. The ultimate implications of this, especially for cosmology are undoubtedly profound but still unknown.

Third is the Holographic Principle. In many ways this is the most surprising ingredient. The non-redundant degrees of freedom that describe a region of space are in some sense on its boundary, not its interior as they would be in field theory. At one per Planck area, there are vastly fewer degrees of freedom than in a field theory, cutoff at the Planck volume. The number of degrees of freedom per unit volume becomes arbitrarily small as the volume gets large. Although the Holographic Principle was regarded with skepticism at first it is now part of the mainstream due to Maldacena's AdS/CFT duality. In this framework the Holographic Principle, Black Hole Complementarity and the IR/UV connection are completely manifest. What is less clear is the dictionary for decoding the CFT hologram.

Finally, the existence of black hole entropy indicates the existence of microscopic degrees of freedom which are not present in the usual Einstein theory of gravity. It does not tell us what they are. String theory does provide a microscopic framework for the use of statistical mechanics. In all cases the entropy of the appropriate string system agrees with the Bekenstein–Hawking entropy. This, if nothing else, provides an existence proof for a consistent microscopic theory of black hole entropy.

The theory of black hole entropy is incomplete. In each case a trick, specific to the particular kind of black object under study, is used to determine the relation between entropy and mass for the specific string-theoretic object that is believed to represent a particular black hole. Then classical general relativity is used to determine the area–mass relation and the Bekenstein–Hawking entropy. In no case do we use string theory directly to compare entropy and area. In this sense the complete universality of the area–entropy relation is still not fully understood.

One very large hole in our understanding of black holes is how to think about the observer who falls through the horizon. Is this important? It is if you are that observer. And in some ways, an observer in a cosmological setting is very much like one behind a horizon. At the time of the writing of this book there are no good ideas about the quantum world behind the horizon. Nor for that matter is there any good idea of how to connect the new paradigm of quantum gravity to cosmology. Hopefully our next book will have more to say about this.

Bibliography

1. Kip S. Thorne, Richard H. Price and Douglas A. MacDonald. Black Holes: The Membrane Paradigm (Yale University Press, 1986).

2. Don N. Page. Information in Black Hole Radiation, hep-th/9306083, Phys. Rev. Lett. 71 (1993) 3743–3746.

3. Leonard Susskind. The World as a Hologram, hep-th/9409089, J. Math. Phys. 36 (1995) 6377–6396.

4. S. Corley and T. Jacobson. Focusing and the Holographic Hypothesis, gr-qc/9602043, Phys. Rev. D53 (1996) 6720–6724.

5. W. Fischler and L. Susskind. Holography and Cosmology, hep-th/9806039.

6. Raphael Bousso. The Holographic Principle, hep-th/0203101, Rev. Mod. Phys. 74 (2002) 825–874.

7. Juan M. Maldacena. The Large N Limit of Superconformal Field Theories and Supergravity, hep-th/9711200, Adv. Theor. Math. Phys. 2 (1998) 231–252; Int. J. Theor. Phys. 38 (1999) 1113–1133.

8. Edward Witten. Anti De Sitter Space and Holography, hep-th/9802150, Adv. Theor. Math. Phys. 2 (1998) 253–291.

9. L. Susskind and Edward Witten. The Holographic Bound in Anti-de Sitter Space, hep-th/9805114.

10. T. Banks, W. Fischler, S.H. Shenker and L. Susskind. M Theory as a Matrix Model: A Conjecture, hep-th/9610043, Phys. Rev. D55 (1997) 5112–5128.

Index

Printed in the United States
By Bookmasters